MEDICAL
INTELLIGENCE
UNIT

GENETIC MECHANISMS IN MULTIPLE ENDOCRINE NEOPLASIA TYPE 2

Barry D. Nelkin
Johns Hopkins University
Johns Hopkins Oncology Center
Baltimore, Maryland, U.S.A.

Springer-Verlag Berlin Heidelberg GmbH

R.G. LANDES COMPANY
AUSTIN

MEDICAL INTELLIGENCE UNIT
GENETIC MECHANISMS IN MULTIPLE ENDOCRINE NEOPLASIA TYPE 2

R.G. LANDES COMPANY
Austin, Texas, U.S.A.

Originally published by Springer-Verlag, Heidelberg, Germany in 1996
Softcover reprint of the hardcover 1st edition 1996

Springer

International ISBN 978-3-662-21950-8

Library of Congress Cataloging-in-Publication Data

Genetic mechanisms in multiple endocrine neoplasia type 2/ {edited by} Barry D. Nelkin.
p. cm. — (Medical intelligence unit)
Includes bibliographical references and index.

ISBN 978-3-662-21950-8 ISBN 978-3-662-21948-5 (eBook)
DOI 10.1007/978-3-662-21948-5

1. Endocrine glands--Tumors--Genetic aspects. 2. Adenomatosis, Familial endocrine--Genetic aspects. I. Nelkin, Barry D., 1951- II. Series.

{DNLM: 1. Multiple Endocrine Neoplasia Type 2--genetics. 2. Thyroid Neoplasms--genetics. 3. Carcinoma. Medullary--genetics. WK 270 G328 1996}
RC280.E55G46 1996
616.99'44042--dc20
DNLM/DLC 96-30948
for Library of Congress CIP

Publisher's Note

R.G. Landes Company publishes six book series: *Medical Intelligence Unit, Molecular Biology Intelligence Unit, Neuroscience Intelligence Unit, Tissue Engineering Intelligence Unit, Biotechnology Intelligence Unit* and *Environmental Intelligence Unit.* The authors of our books are acknowledged leaders in their fields and the topics are unique. Almost without exception, no other similar books exist on these topics.

Our goal is to publish books in important and rapidly changing areas of bioscience and environment for sophisticated researchers and clinicians. To achieve this goal, we have accelerated our publishing program to conform to the fast pace in which information grows in bioscience. Most of our books are published within 90 to 120 days of receipt of the manuscript. We would like to thank our readers for their continuing interest and welcome any comments or suggestions they may have for future books.

Shyamali Ghosh
Publications Director
R.G. Landes Company

CONTENTS

EDITOR

Barry D. Nelkin, Ph.D.
The Oncology Center
Johns Hopkins University
School of Medicine
424 N. Bond Street
Baltimore, Maryland, U.S.A.
Chapters 8, 10

CONTRIBUTORS

Douglas W. Ball, M.D.
Division of Endocrinology and Metabolism and Oncology Center
Johns Hopkins University School of Medicine
Baltimore, Maryland, U.S.A.
Chapter 1

Francesca Carlomagno, M.D.
Centro di Endocrinologia ed Oncologia Sperimentale del CNR
Facoltà di Medicina e Chirurgia
Università di Napoli "Federico II"
Naples, Italy
Chapter 3

Nina A. Dathan
Centro di Endocrinologia ed Oncologia Sperimentale del CNR
Facoltà di Medicina e Chirurgia
Università di Napoli "Federico II"
Naples, Italy
Chapter 3

Ronald A. DeLellis, M.D.
Department of Pathology
Tufts University School of Medicine and New England Medical Center
Boston, Massachusetts, U.S.A.
Chapter 6

Pier Paolo Di Fiore, M.D., Ph.D.
Laboratory of Cellular and Molecular Biology
National Cancer Institute
National Institutes of Health
Bethesda, Maryland, U.S.A.
Chapter 3

Alfredo Fusco, M.D.
Dipartimento di Medicina Sperimentale e Clinica
Facoltà di Medicina e Chirurgia di Catanzaro
Università di Reggio Calabria
Catanzaro, Italy
Chapter 3

Gary Landreth, Ph.D.
Department of Neurology and Neurosciences
Alzheimer Research Laboratory
Case Western Reserve University School of Medicine
Cleveland, Ohio, U.S.A.
Chapter 4

Thomas M. Lanigan
Molecular Biology Program
University of Iowa
Iowa City, Iowa, U.S.A.
Chapter 7

Jeffrey F. Moley, M.D.
Washington University School of Medicine
and St. Louis Veteran's Administration Medical Center
St. Louis, Missouri, U.S.A.
Chapter 9

Michael P. Myers, Ph.D.
Department of Neurology and Neurosciences
Alzheimer Research Laboratory
Case Western Reserve University School of Medicine
Cleveland, Ohio, U.S.A.
Chapter 4

M.A. Pierotti, Ph.D
Oncologia Sperimentale A
Istituto Nazionale Tumori
Milan, Italy
Chapter 2

B.A.J. Ponder, Ph.D., FRCP
CRC Human Cancer Genetics Research Group
Addenbrooke's Hospital
Hills Road, Cambridge, United Kingdom
Chapter 2

Andrew F. Russo, Ph.D.
Department of Physiology and Biophysics
Molecular Biology Program
University of Iowa
Iowa City, Iowa, U.S.A.
Chapter 7

Massimo Santoro, M.D., Ph.D.
Centro di Endocrinologia ed Oncologia Sperimentale del CNR
Facoltà di Medicina e Chirurgia
Università di Napoli "Federico II"
Naples, Italy
Chapter 3

Kenneth D. Swanson
Department of Neurology and Neurosciences
Alzheimer Research Laboratory
Case Western Reserve University School of Medicine
Cleveland, Ohio, U.S.A.
Chapter 4

Arthur S. Tischler, M.D.
Department of Pathology
Tufts University School of Medicine and New England Medical Center
Boston, Massachusetts, U.S.A.
Chapter 6

Giancarlo Vecchio, M.D., Ph.D.
Centro di Endocrinologia ed Oncologia Sperimentale del CNR
Facoltà di Medicina e Chirurgia
Università di Napoli "Federico II"
Naples, Italy
Chapter 3

Kristine S. Vogel, Ph.D.
Center for Developmental Biology
University of Texas Southwestern Medical Center
Dallas, Texas, U.S.A.
Chapter 5

PREFACE

From a variety of scientific perspectives, the multiple endocrine neoplasia, type 2 (MEN 2) syndromes provide a setting for the study of a fascinating array of important biological questions. In turn, our understanding of basic biological processes, such as signal transduction and development, has rapidly augmented our insight into the biological behavior of the cancers of MEN 2. Finally, activating mutations in the ret tyrosine kinase gene have been shown to underlie MEN 2. These mutations represent the first demonstration of an activating mutation in an oncogene in a hereditary cancer syndrome; this finding has served to focus our attention regarding the specific molecular mechanisms of tumor development in MEN 2.

In order to understand the biology of MEN 2 syndromes, it is necessary to take an integrated approach. Mutations and alterations in gene expression and signal transduction must be placed in context of their effects on the biology of the thyroid C-cells and adrenal chromaffin cells which are the progenitors of medullary thyroid carcinoma and pheochromocytoma. This will require an understanding of the developmental, cell and molecular biology of these cells, as well as an understanding of the function of normal or mutated ret proteins. This book is a first attempt to bring together these topics. In chapter 1, Ball discusses the clinical aspects of MEN 2, including the impact on clinical management of MEN 2. In chapter 2, Ponder and Pierotti address the spectrum of ret mutations in MEN 2, and their implications for tumorigenesis. In chapter 3, Fusco et al examine the signal transduction pathways which may be affected by ret activation. In chapter 4, Myers et al review the signal transduction pathways which control differentiation in the well-characterized pheochromocytoma cell line, PC12. Chapters 5 and 6 discuss the normal and abnormal development of adrenal chromaffin cells from the neural crest. Vogel (chapter 5) details the normal migration of neural crest cells to the adrenal medulla, their commitment to the sympathoadrenal lineage and their differentiation into neuronal or chromaffin cells. Tischler and DeLellis (chapter 6) examine how, in MEN 2 and in rodent models of pheochromocytoma, the adrenal medulla undergoes hyperplasia, culminating in tumor development. Russo and Lanigan (chapter 7) review the neuronal properties of C-cells, and discuss possible effects on this differentiation during MTC tumorigenesis. Nelkin (chapter 8) and Moley (chapter 9) discuss possible steps in tumor progression in MTC and pheochromocytoma.

Acknowledgment

Clearly, the contributing authors deserve the credit for this work. In addition, I thank my colleagues at the Johns Hopkins Oncology Center; their insights have provided some needed direction in my research.

The research in my laboratory is currently supported by grants from the National Cancer Institute and the American Cancer Society.

CHAPTER 1

Clinical Manifestations of Multiple Endocrine Neoplasia Type 2

Douglas W. Ball

Though uncommon in the general population, the multiple endocrine Neoplasia type 2 (MEN 2) syndromes are noteworthy for their distinctive genetic, developmental and biochemical features, and their unique clinical associations. Elucidation of the genetic basis of the MEN 2 syndromes has spawned substantial improvements in diagnosis which in turn have led to more prompt and appropriate treatment of the associated endocrine tumors. In the foreseeable future, greater understanding of the molecular pathogenesis of these tumors is likely to allow a more rational targeting of anti-tumor therapy. This introductory chapter reviews the natural history and summarizes current approaches to diagnosis and management of patients with the MEN 2 syndromes.

HISTORY AND CLASSIFICATION

Medullary thyroid carcinoma (MTC), the cardinal lesion of MEN 2, was recognized as a discrete histologic type of thyroid cancer by Hazard in 1959.[1] Prior to this time, many MTC patients were classified as having poorly differentiated or anaplastic thyroid cancers. A clinical syndrome ultimately classified as MEN 2 was first identified by Sipple who reported a high incidence of

Genetic Mechanisms in Multiple Endocrine Neoplasia Type 2, edited by Barry D. Nelkin.

thyroid cancer in patients with pheochromocytomas.[2] The nature of the Sipple syndrome was further clarified by other investigators who recognized that MTC was inherited in an autosomal dominant pattern with pheochromocytoma, along with hyperparathyroidism or less frequently, multiple mucosal neuromas.[3-6] The principal secreted product of MTC, calcitonin, was identified as a circulating hypocalcemic hormone by Copp and colleagues in 1962.[7] Calcitonin production by the thyroid parafollicular cells was subsequently demonstrated by Foster.[8] Three subsequent studies together provided the basis of our contemporary concept of MTC histogenesis. Williams first demonstrated that MTC arises from the parafollicular C-cells,[9] which were found to originate in the fetal neural crest,[10,11] in distinction from the endoderm-derived thyroid epithelium. Melvin and Tashjian showed in 1968 that MTC tumors produce and secrete calcitonin.[12] Critical steps in the understanding of MEN 2 genetics were the discovery of linkage to chromosome 10 loci in 1987,[13,14] and the identification of ret protooncogene mutations in these disorders by Mulligan and colleagues[15] and other groups[16,17] in 1993.

Medullary thyroid carcinoma is an uncommon thyroid malignancy, accounting for 3 to 5% of thyroid cancers.[18,19] Unlike the common epithelial forms of thyroid cancer, MTC is inherited in approximately 20% of cases as an autosomal dominant trait in one of three distinct syndromes (Table 1.1). The term MEN 2 was originally proposed by Steiner[20] to distinguish the syndrome of MTC, pheochromocytoma and hyperparathyroidism from Wermer's syndrome or MEN 1 (parathyroid, pituitary, and pancreatic islet tumors). Multiple endocrine neoplasia type 2A or MEN 2A is now the preferred designation for the syndrome defined by Steiner.[20] MEN 2B refers to a distinct syndrome of MTC, pheochromocytoma, multiple mucosal and intestinal ganglioneuromas, and a Marfanoid body habitus, with less than 25% of the prevalence of MEN 2A. A syndrome of familial MTC in the absence of other endocrinopathy, termed FMTC, was characterized by Farndon and colleagues.[22] Extensive molecular genetic studies, detailed in succeeding chapters, have confirmed that all three of these MEN 2-related syndromes stem from characteristic mutations of the ret tyrosine kinase receptor gene. In addition, two interesting clinical variants of MEN 2A have been described. MEN 2A with cutaneous lichen amyloidosis (CLA) consists of a unique

Table 1.1. Classification of medullary thyroid carcinoma and the MEN 2 syndromes

Type	Associated Lesions	ret Gene Mutation	Clinical Behavior
Sporadic	none	Somatic (~33% of tumors) Tyrosine Kinase Domain	Intermediate
FMTC	none	Germline Extracellular Domain	Less aggressive
MEN 2A	Pheochromocytoma Hyperparathyroidism	Germline Extracellular Domain	Intermediate
MEN 2B	Pheochromocytoma Ganglioneuromas Marfanoid Habitus	Germline Tyrosine Kinase Domain	Aggressive

From Ref. 80, with permission of Lippincott-Raven Publishers.

pruritic plaque-like rash located over the scapular area in addition to typical manifestations of MEN 2A.[23,24] MEN 2A with concurrent Hirschsprung's disease, has been linked to a specific subset of ret gene mutations associated with MEN 2A.[25] Individuals from these kindreds may exhibit normal colonic function or varying degrees of colonic aganglionosis.

CLINICAL PRESENTATION AND DIAGNOSIS: MTC

Patients with the more common sporadic form of MTC most often present with a painless unilateral thyroid nodule, typically in the fifth or sixth decade of life. Earlier presentations in the twenties or thirties or are not uncommon in sporadic MTC but should raise a strong suspicion of inherited disease.[26] There is a slight female preponderance in sporadic MTC of approximately 1.5:1; no gender imbalance is seen in MEN 2. The age-specific penetrance of C-cell disease in MEN 2A gene carriers has been carefully studied by Ponder and associates. At age 35, the likelihood of a clinical presentation with MTC is only 25% for obligate gene carriers. Even at age 70, the likelihood is only approximately 60%.[26] The incomplete penetrance of clinically recognized MTC provides one explanation for the phenomenon of heritable ret gene mutations in individuals with no apparent family history, discussed below. A more sensitive look at the age-specific penetrance of C-cell disease in MEN 2A was afforded by serial biochemical screening using the calcitonin secretogogues, calcium and pentagastrin. Approximately 65% of obligate gene carriers exhibited calcitonin hypersecretion at age 20. At age 35, 95% of gene carriers had a positive stimulation test.[26] Thus, the penetrance of C-cell disease, employing more sensitive indicators, is extremely high. With prospective germline DNA testing for ret mutations, both C-cell hyperplasia and microscopic MTC can be readily diagnosed prior to conversion to a positive biochemical screening test.[27,28]

The progression of MTC is distinguished by several well-characterized stages of tumor formation. In the MEN 2 syndromes, the initial manifestation is diffused C-cell hyperplasia, followed by outgrowth of multiple foci of microscopic carcinoma. Each individual cancer represents a separate clone of tumor cells that has presumably arisen from a single hyperplastic C-cell.[29] In MEN 2B, C-cell hyperplasia may be detectable at birth and frank carcinomas may appear as early as the first year of life.[30] A subset of MEN 2B kindreds follow a relatively less aggressive course,[31] although they harbor ret tyrosine kinase domain mutations at Met 918 that appear indistinguishable from more typical MEN 2B families. In MEN 2A, progression from C-cell hyperplasia to microscopic MTC occurs at a variable rate, beginning in the second half of the first decade of life. Previously, the presence of significant C-cell hyperplasia in a patient frequently was assumed to be a hallmark of MEN 2. With the advent of widespread testing of MEN 2 kindreds for ret gene mutations, it has become increasingly apparent that some obligate non-carriers of the ret mutation may test positively by biochemical stimulation tests for C-cell abnormalities and have thyroidectomy specimens that meet rigorous definitions of C-cell hyperplasia.[28] Although false negative ret testing is a possible explanation, follow-up testing for evidence of recurrent MTC or pheochromocytoma has thus far been negative in these patients. C-cell hyperplasia has been reported in conjunction with hyperthyroidism, lymphocytic thyroiditis, papillary and follicular thyroid cancer, and in unselected patients at autopsy.[32] Thus, the prevalence of C-cell hyperplasia in the overall population is probably greater than initially suggested.

At the time of presentation with a palpable thyroid nodule, greater than 50% of MTC patients have detectable cervical lymphadenopathy, whereas distant metastasis to lung, liver or bone are identifiable in approximately 10-15% of cases at presentation. Typical patterns of MTC regional lymphadenopathy and the impact of systematic lymph node exploration have been reviewed,[33-35] although incorporation of these data into widespread surgical practice has been incomplete in the U.S. Of great clinical importance is the observation that even the smallest clinically identifiable tumors can produce lymph node metastasis. Locally invasive lesions may present with dysphagia, neck pain, or especially hoarseness, related to recurrent laryngeal nerve injury. Bulky tumors, frequently in the setting of liver metastases, may be associated with diarrhea and flushing. Either MTC or pheochromocytoma may ectopically secrete ACTH, leading to Cushing's syndrome.[36,37]

Markedly elevated blood levels of calcitonin (frequently 20- to 1,000-fold above normal) are typical in patients with clinically evident MTC. Apart from patients with early inherited disease, borderline elevations of calcitonin usually are not associated with MTC. Instead, a variety of conditions including inflammatory and neoplastic lung disease, pancreatic and gastrointestinal tumors, renal failure as well as non-specific physiologic stimuli can be associated with low-grade calcitonin elevations.[38,39] Although acute infusions of calcitonin result in transient hypocalcemia and compensatory increases in parathyroid hormone, chronic calcitonin excess in MTC does not result in either hypocalcemia or compensatory hyperparathyroidism. The role of calcitonin in MTC-related diarrhea and flushing is controversial, but other mediators including vasoactive intestinal peptide, serotonin metabolites and prostaglandins have been implicated more frequently.

As an adjunct to basal calcitonin determinations, many centers employ calcitonin provocative testing, using either pentagastrin (5 µg/kg) alone or pentagastrin plus calcium (2 mg elemental Ca^{++}/kg) as secretogogues.[40] Using commonly employed protocols and a sensitive calcitonin assay (5 pg/ml detection limit), typical peak normal values are less than 50 pg/ml (for pentagastrin alone) or 350 (for pentagastrin plus calcium in males or 95 in females). As discussed in later chapters, ret mutation testing has largely superseded calcitonin provocative testing for screening asymptomatic subjects in known kindreds or for determining the heritability of apparently sporadic MTC. The principal uses of calcitonin provocative testing at this point are: (1) to evaluate the extent of disease in subjects testing positive in genetic screening; (2) to evaluate for early recurrence in MTC patients who have undergone thyroidectomy; and (3) to assist in the family screening of the minority of MEN 2A and FMTC kindreds who lack a detectable ret gene mutation. Successful thyroidectomy in MTC patients, exemplified by children identified as ret mutation carriers in early pre-symptomatic screening, usually results in a stimulated calcitonin level that is near to the assay limit of detection, typically less than 10 pg/ml.[27] While the prognostic significance of somewhat higher post-operative calcitonin values has not been precisely determined, it is likely that stimulated values that rise progressively on follow-up testing represent low-grade persistent disease, even when these values lie nominally within the normal range.

In addition to calcitonin, MTC tumors produce an alternatively spliced product of the calcitonin gene transcript, calcitonin gene-related peptide (CGRP).[41] CGRP is a potent vasodilator;[42] its role, if any, in MTC pathophysiology, is unclear. Although both hormones may circulate at high levels in the blood of MTC patients, CGRP basal and stimulated levels are more variable than are levels of calcitonin.[43,44] Katacalcin, a 21 amino acid carboxyl terminal peptide derived by post-translational processing of the calcitonin pro-peptide is co-secreted with calcitonin and has unknown biological function.[45]

Diagnosis of macroscopic MTC is based on the clinical detection of a thyroid nodule and on fine needle aspiration cytology. Routinely stained aspirates reveal a highly cellular, often heterogeneous pattern that lacks typical papillary or follicular features and is depleted of colloid.[32] Performance of calcitonin immunohistochemistry, combined with a serum calcitonin determination, usually confirms the diagnosis. Following initial detection, a useful pre-operative staging workup includes CT, MRI or ultrasound of the neck to identify lymphadenopathy. CT or MRI of the chest are used to screen for mediastinal lymphadenopathy or lung parenchymal metastases. Detection of liver metastasis remains a significant challenge in MTC imaging, due to the diffuse, miliary pattern of these tumor deposits. Although both selective venous sampling and laparoscopic liver biopsy have been used successfully to identify early liver metastasis, neither procedure is frequently performed prior to initial thyroid surgery. It is prudent for most patients undergoing surgery for MTC to have a urinary catecholamine determination in order to rule out occult pheochromocytoma.

A final task in MTC diagnosis is the evaluation for heritable disease. The indications and techniques for screening for heritable MTC both in known kindreds and in patients with no apparent family history are discussed in a later section.

CLINICAL PRESENTATION AND DIAGNOSIS: PHEOCHROMOCYTOMA

Pheochromocytoma, a catecholamine-producing tumor derived from adrenal medullary chromaffin cells, is a rare cause of hypertension in the general population, accounting for <0.1% of cases. The clinical importance of this entity is magnified by the severe and potentially life-threatening consequences of undetected pheochromocytoma. In historical analyses of MEN 2 kindreds, sudden unexplained death, premature stroke and myocardial infarction related to uncontrolled hypertension were principal causes of early death, even exceeding the mortality associated with medullary thyroid cancer.[46] Fortunately, genetic screening together with periodic biochemical testing for catecholamine excess and modern imaging and surgical techniques have effectively minimized the mortality attributable to pheochromocytoma in MEN 2. Indeed, a recent clinical series reported less morbidity and mortality stemming from pheochromocytoma per se than from complications related to decompensation and inadequate steroid replacement in adrenalectomized patients.[47]

In kindreds with inherited MTC, the presence of pheochromocytoma is the major determinant of a diagnosis of MEN 2A versus FMTC. The lifetime penetrance of pheochromocytoma across a large number of MEN 2A kindreds is approximately 50%. Individual kindreds vary quite widely with lifetime prevalence rates ranging from 10 to 90%. Because of the significant overlap in ret mutations observed in MEN 2A and FMTC, the molecular determinants of pheochromocytoma in MEN 2 remain somewhat obscure. Overall, mutations at Cys 634 are far more common in MEN 2A (87.1%) than in FMTC (30%).[48] In addition, some kindreds originally classified as FMTC ultimately prove to have a low prevalence of pheochromocytoma.

The clinical manifestations of pheochromocytoma in sporadic and familial cases are generally similar, with several important differences noted below. Individually, the symptoms of pheochromocytoma are relatively non-specific. However, the clustering of these symptoms, particularly in paroxysms lasting minutes to hours, can strongly suggest the diagnosis. The majority of paroxysms occur spontaneously when the patient is calm and at rest. Paroxysms occasionally may be precipitated by movements that result in abdominal pressure, by strenuous exertion, or rarely by urination or defecation. Emotional triggers do not appear to play a significant role. In order of decreasing frequency, headache (80%), diaphoresis (71%), chest palpitation (64%) and nervousness or anxiety are the most typical symptoms.[49] Headache is frequently severe, constant and diffuse. The development of a new or worsening headache pattern in MEN 2 patients with intact adrenal glands should always prompt a biochemical search for catecholamine excess.

Hypertension in pheochromocytoma may be sustained (60%), paroxysmal (25%) or absent (15%).[50] Even in patients with sustained hypertension, blood pressure may be highly labile. Hypertension in these patients is frequently refractory to routine treatment with diuretics and beta-adrenergic blockers, but is more sensitive to alpha-adrenergic blockers (labetalol, phenoxybenzamine, prazosin). Complications of acute hypertensive crises include retinal hemorrhages, proteinuria, hematuria and cardiac ischemia. High circulating catecholamine levels in pheochromocytoma typically result in vasoconstriction, diminished plasma volume and episodes of orthostatic hypotension. Chronic metabolic complications include glucose intolerance and occasionally hypercalcemia.

Catecholamine-secreting tumors may stem directly from the adrenal medulla or from extramedullary chromaffin cells (paragangliomas or "extra-adrenal pheochromocytomas"). In sporadic cases, more than 80% of tumors are located unilaterally within the adrenal gland. Extra-adrenal locations generally follow the distribution of sympathetic ganglia including the neck, posterior mediastinum and especially the para-aortic sympathetic chains terminating in the organ of Zuckerkandl. Rare locations include the urinary bladder, pericardium, carotid body and cranial nerves. The estimation of malignant potential, based on local invasiveness and distant metastasis ranges from 10-25% of tumors.

Familial pheochromocytomas comprise approximately 10% of the total and provide unique clinical challenges. The principal familial pheochromocytoma syndromes are Von-Hippel Lindau disease (VHL), MEN 2A, MEN 2B and neurofibromatosis type I

(NF1). Isolated familial pheochromocytoma is a questionable disease entity. Many families originally identified with this syndrome appear, after more detailed study, to have unrecognized VHL.[51] Other autosomal dominant disorders with a rare incidence of pheochromocytoma include the neuroectodermal syndromes tuberous sclerosis and Sturge-Weber disease[52] and a syndrome of familial extra-adrenal pheochromocytomas.[53] The actual prevalence of familial disease among unselected patients presenting with pheochromocytoma is uncertain. Neumann and colleagues[51] reported 16 occult cases of VHL and 3 cases of unsuspected MEN 2A among 82 unselected pheochromocytoma cases. The unexpectedly high prevalence of these disorders may reflect a referral bias in this tertiary care population. Occasionally, pheochromocytoma may be the first symptomatic lesion in VHL. In such patients, asymptomatic retinal angiomatosis is frequently present and provides the most readily available screening tool. Interestingly, VHL families with a high incidence of pheochromocytoma have a corresponding lower incidence of renal cell carcinoma[51] and a heavy bias to a single VHL gene mutation site at Arg238.[54]

Familial pheochromocytomas in general, and MEN 2A-associated pheochromocytomas in particular, vary in several important clinical parameters compared to sporadic tumors: age of incidence, bilaterality, occurrence of malignancy and catecholamine metabolism. First, the median age of presentation is significantly younger: 37 in MEN 2A and 27 in VHL, versus 46 in sporadic pheochromocytoma.[51] To some extent, pre-symptomatic screening is responsible for earlier detection in both MEN 2 and in VHL. Data from the European MEN 2 (EUROMEN) study group comprise 274 MEN 2A and 26 MEN 2B individuals studied between 1966 and 1982.[46] In this series, the median age of diagnosis of pheochromocytoma was 36 years (range 14-68) in MEN 2A and 32 years (range 15-41) in MEN 2B. Pheochromocytoma was detected first in 28% of cases, synchronously with MTC in 32% and after MTC in 40% of cases, reflecting a case mix that included both symptomatic detection and pre-symptomatic biochemical screening. Only 39% of the MEN 2A pheochromocytoma patients had symptoms, including hypertension, at the time of diagnosis.

In MEN 2A family screening programs that employ annual testing with urinary catecholamines or metabolites (metanephrines), the great majority of affected patients can be detected prior to the onset of hypertension. While virtually all MEN 2A patients with frankly abnormal catecholamine levels will exhibit abnormalities on abdominal MRI or CT scanning, the converse is not always true.[55] Compared to sporadic tumors, pheochromocytomas in MEN 2 produce an excess of epinephrine relative to norepinephrine.[56] Activity of PNMT (phenylethanolamine-N-methyltransferase) appears to be the major determinant of epinephrine versus norepinephrine secretion, and is a critical differentiation marker in developing adrenomedullary precursor cells.[57,58]

The bilateral nature of familial pheochromocytomas has been well documented in several case series. As described later in this monograph, pheochromocytomas in MEN 2 develop on a background of polyclonal adrenomedullary hyperplasia.[59] Depending in part on the extent of clinical follow-up and variations in surgical practice, 68-75% of MEN 2A patients with pheochromocytoma ultimately prove to have bilateral tumors.[46,51] Multifocality in MEN 2B is equally as common. The interval between detection of the first and second pheochromocytoma can vary substantially from synchronous presentations to gaps of more than 20 years. Thus, patients undergoing unilateral adrenalectomy should have annual repeat screening, extending indefinitely. Extra-adrenal pheochromocytoma (paraganglioma) is uncommon in MEN 2 patients (< 10%), though not infrequent in VHL (30%).[51] The true incidence of malignant pheochromocytoma in MEN 2A is somewhat controversial. Several published reports[51,55] suggest an extreme paucity of such cases. In contrast, the EUROMEN study indicated a 4% rate of malignancy with 4 of 300 MEN 2 patients dying of metastatic pheochromocytoma during the period of follow-up. This 4% rate appears to be significantly lower than the 10-25% rate of metastasis or invasion seen in sporadic pheochromocytoma.

HYPERPARATHYROIDISM IN MEN 2

Hyperparathyroidism occurs in approximately 10-15% of individuals with the MEN 2A syndrome.[60] As in the case of pheochromocytoma, genotype is an incomplete predictor of the prevalence of hyperparathyroidism among different MEN 2A kindreds. A bias toward a single ret gene point mutation (634 TGC (Cys)->CGC (Arg)) has been reported.[61] MEN 2B is not associated with a comparable increase in hyperparathyroidism. Parathyroid histology in MEN 2A consists of a background of parathyroid hyperplasia with superimposed adenomas.[62] From a clinical standpoint, parathyroid disease in MEN 2A is often limited to asymptomatic hypercalcemia, although more severe manifestations such as fatigue, weakness, kidney stones and osteitis fibrosa cystica may occur. Diagnosis with plasma calcium (corrected for albumin) and a PTH level is generally straightforward. Although pre-operative localization with ultrasonography, neck CT or scintigraphic scans is frequently employed, the surgeon, as in the case of MEN 1, is obligated to examine each parathyroid gland in these patients. Surgical management typically follows two approaches: (1) three and half gland parathyroidectomy with identification of the remaining fragment with surgical clips; or (2) total parathyroidectomy with forearm re-implantation. Disease recurrences, as in MEN 1, present a significant surgical challenge.

INTEGRATED APPROACH TO PRE-SYMPTOMATIC SCREENING FOR MEN 2

There has been a dramatic improvement in survival in families with MEN 2A, FMTC and to a lesser extent MEN 2B over the last 35 years. This improvement in outcome is largely attributable to the success of family screening programs, first employing provocative biochemical tests[63] and more recently linkage-based and mutation-based genetic tests.[27,28] The importance of early detection is underscored by several observations. First, individuals who have pre-existing calcitonin hypersecretion at the time of the initial biochemical screening have a significantly higher rate of post-operative MTC recurrence than do individuals who developed calcitonin hypersecretion after several years of negative provocative testing.[63] Secondly, microscopic carcinomas can be found in subjects identified by genetic screening who have normal calcitonin provocative tests. Third, metastatic MTC in MEN 2A apparently has not been reported prior to age 6, but has been demonstrated occasionally in subjects between ages 6 and 10. The principal goal of family screening programs is therefore to identify affected children at an age at which metastatic MTC is unlikely, but thyroidectomy can be accomplished in a safe and non-traumatic fashion.

A widely accepted screening program for known MEN 2A kindreds is to initiate testing for ret gene mutations at approximately ages 4 to 6. In the case of MEN 2B, recognition of the characteristic oral mucosal neuroma phenotype is generally possible in the first two years of life. Identification of the typical codon 918 Met -> Thr mutation confirms the diagnosis.

A variety of testing procedures have been proposed, all incorporating PCR amplification of germline DNA. Typical amplification schemes target ret exons 10, 11 and 16, employing either direct automated DNA sequencing, or allelic mismatch detection procedures, including SSCP and DGGE, in order to identify mutations. Once a prototype family mutation is identified, restriction analysis often provides a simple means for screening additional members and a useful adjunct to sequencing techniques. Currently, a small number of commercial and research laboratories routinely perform these assays.

The diagnostic accuracy of family screening programs is potentially limited by several important factors. As discussed in chapter 2, to date, 3-5% of MEN 2A kindreds studied and 13-20% of FMTC kindreds have revealed no detectable ret gene mutations despite fairly intensive study.[48,61] An additional rare mutation site on ret exon 13 only accounts for a small percentage of the remainder.[64] Molecular characterization of these remaining kindreds will shed additional light on the mechanisms of MEN 2 initiation. Once a characteristic mutation is discovered for a given family, the rate of false negative screening results is extremely low. An additional source of diagnostic error is allele-specific amplification related to polymorphisms that may render one allele non-amplifiable with a given primer set.[65] If the polymorphic allele happens to harbor the mutation, then it would be under-represented in the resulting PCR product. Another possible source of error is the mislabeling of samples. In general, a repeat determination from a separate blood drawing should be contemplated prior to undertaking surgery on the basis of a DNA test result.

ROLE OF DNA SCREENING IN APPARENTLY SPORADIC MTC

Based on earlier studies by Ponder and colleagues, it is clear that a small fraction of MTC patients with no known family history of MEN 2-related tumors actually have the potential to pass the disease to their offspring. Based on provocative calcitonin testing in children of such affected adults, it was estimated that 6-10% of apparently sporadic patients actually harbor germline mutations.[26]

This prediction has recently been borne out in studies by Gagel and others, that have revealed typical MEN 2A-like germline ret mutations.[66] Small family size, incorrect assumptions regarding paternity, the incomplete penetrance of clinically recognizable MTC and occasional de novo germline mutations may all contribute to this phenomenon. Although no broad consensus for screening in these patients has yet been reached, it appears prudent for patients with MTC and no apparent family history to undergo germline ret gene analysis to rule out occult heritable disease. In addition, approximately 30-50% of sporadic MTC tumors exhibit somatic (acquired) ret mutations of codon 918, identical to those seen in MEN 2B.[17,67,68] The prognostic value of this mutation in sporadic MTC is currently under investigation. For patients presenting with apparently sporadic pheochromocytoma, screening recommendations are less clear-cut. Some centers recommend calcitonin determinations plus slit lamp examinations for evidence of occult VHL in all of such patients.[51]

TREATMENT STRATEGIES AND PROGNOSIS IN MEN 2

Based on the poor success of interventions directed at disseminated MTC, treatment strategies revolve around early, pre-symptomatic detection of ret gene mutation carriers followed by prophylactic total or near total thyroidectomy. In some centers, total parathyroidectomy with forearm re-implantation is performed on subjects considered at high risk for future hyperparathyroidism. For patients presenting with a macroscopic thyroid nodule and no clinical evidence of extra-thyroidal disease, the procedure of choice is generally total thyroidectomy and lymph node dissection of the central compartment from the hyoid bone to the innominate veins. Patients with known or suspected inherited MTC, presenting with macroscopic disease, should be considered candidates for bilateral neck exploration. The mid-jugular nodes should be sampled. If they are positive for tumor, a modified neck dissection is frequently indicated with sparing of the sternocleidomastoid muscle, jugular vein and accessory nerve. Radical neck dissection has not been shown to improve prognosis.

In practical terms, a surgical cure of MTC is defined as a normal post-operative calcitonin provocative test result that persists over time. Overall, the rate of persistent hypercalcitoninemia is nearly 50% for patients with non-palpable macroscopic MTC and greater than 80% for patients presenting with palpable MTC. Many of these patients have no abnormalities on neck and chest CT scans, suggesting the presence of microscopic lymph node deposits or distant occult disease (such as liver metastasis). In the absence of extensive adenopathy or overt organ involvement, the clinical course of persistent MTC is usually marked by very gradual progression. A Mayo Clinic retrospective series indicated excellent long-term survival in patients with persistent calcitonin elevation but no other evident disease following primary surgery (86% overall survival at 10 years).[69]

Given the difficulty of eradicating macroscopic MTC with primary surgery, several groups have employed microdissection techniques in repeat operations with a curative intent. Tisell and associates reported the normalization of post-operative calcitonin levels in 4 of 11 patients.[70] More recently, Moley and colleagues reported that 9 of 32 patients undergoing re-exploration for persistent hypercalcitoninemia had initial post-operative stimulated calcitonin values less than 160 pg/ml.[71] Capsular invasion in the primary surgery specimen was a significant predicator of an unsuccessful re-operation. Other centers have reported lower rates of post-operative calcitonin normalization.[34] It is currently unclear how many of those apparently responded to re-operation will translate into the long-term surgical cures. In addition to re-operation in patients with minimal residual disease, there is a definite role for palliative de-bulking surgery in patients with advanced MTC.

Non-surgical treatment approaches for MTC have generally met with limited success. Although most investigators agree that MTC is not generally radiosensitive,[72,73] external beam radiotherapy of bone metastases can provide useful palliation. A number of small-scale chemotherapy trials in MTC have recently been reviewed, with some regimens providing up to a 30% partial response rate but few if any durable, complete responses.[74,75]

Treatment strategies for pheochromocytoma in MEN 2 have recently been reviewed.[55] Based on the significant morbidity and rare mortality attributed to Addisonian crisis in patients who have undergone total adrenalectomy, most centers currently only remove adrenal glands that are enlarged on routine imaging studies or at surgery. This approach appears to be justified by the extremely low frequency of hypertensive emergencies in patients who undergo unilateral adrenalectomy and then are followed expectantly for recurrence.[47] Recently, laparoscopic techniques have been successfully applied to pheochromocytoma, providing a less invasive approach to this problem in MEN 2.[76]

Although patients with MEN 2A have a significantly better overall prognosis than patients with sporadic MTC, these differences appear to become non-significant when clinical stage is controlled.[73] In smaller patient series, FMTC appears to carry a significantly better prognosis than MEN 2A, whereas MEN 2B is clearly worse. Other clinical factors reported to have predictive value include local invasiveness (67% 10 year survival for patients with capsular invasion versus 92% for patients with an intact capsule),[77] calcitonin immunostaining heterogeneity (60% survival for patients with fewer than 10 % of cells staining positively versus 87% if greater than 50% of cells stain positively),[77] primary tumor size[35] and a rising CEA level.[78,79]

SUMMARY

In conclusion, the MEN 2 syndromes provide a paradigm for the potential benefits of DNA-based diagnostic strategies in the management of an inherited disease. From a clinical perspective, the principal challenges now focus on the treatment of persistent or recurrent MTC. Basic studies in molecular genetics, developmental and cell biology, such as those outlined in the following chapters, are critical to a more comprehensive understanding of the pathogenesis of MEN 2-associated neoplasms.

REFERENCES

1. Hazard JB, Hawk WA, Crile G Jr. Medullary (solid) carcinoma of the thyroid: a clinicopathologic entity. J Clin Endocrinol Metab 1959; 19:152-61.
2. Sipple JH. The association of pheochromocytoma with carcinoma of the thyroid gland. Am J Med 1961; 31:163-6.
3. Cushman P Jr. Familial endocrine tumors. Report of two unrelated kindreds affected with pheochromocytoma, one also with multiple thyroid carcinomas. Am J Med 1962; 32:352-60.
4. Manning PC Jr, Molnar GD, Black BM et al. Pheochromocytoma, hyperparathyroidism, and thyroid carcinoma occurring coincidentally. N Engl J Med 1963; 268:68-72.
5. Nourok DS. Familial pheochromocytoma and thyroid carcinoma. Ann Intern Med 1964; 60:1028.
6. Schimke RN, Hartmann WH, Prout TE et al. Syndrome of bilateral pheochromocytoma, medullary thyroid carcinoma, and mul-

tiple neuromas. N Engl J Med 1968; 279:1-17.
7. Copp DH, Cameron EC, Cheney BA et al. Evidence for calcitonin: a new hormone from the parathyroid that lowers blood calcium. Endocrinology 1962; 70:638-49.
8. Foster GV, MacIntyre I, Pearse AGE. Calcitonin production and the mitochondrion-rich cells of the dog thyroid. Nature 1964; 203:1029-30.
9. Williams ED. Histogenesis of medullary carcinoma of the thyroid. J Clin Pathol 1966; 19:114-8.
10. LeDouarin N, Le Lievre C. Demonstration de l'origine neurale des cellules a calcitonine du corps ultimobranchial chez l'embryon de poulet. Comptes Rendues Seances Acad Sci D Paris 1970; 270:2857-60.
11. Weston, JA. The regulation of normal and abnormal neural crest development. Adv Neurol 1981; 29:77-95.
12. Melvin KEW, Tashjian AH Jr. The syndrome of excessive thyrocalcitonin produced by medullary carcinoma of the thyroid. Proc Natl Acad Sci USA 1968; 59:1261-2.
13. Mathew CGP, Chin KS Easton DF et al. A linked genetic marker for multiple endocrine neoplasia type 2A on chromosome 10. Nature 1987; 328:527-8.
14. Simpson NE, Kidd KK, Goodfellow PJ et al. Assignment of multiple endocrine neoplasia type 2A to chromosome 10 by linkage. Nature 1987; 328:528-9.
15. Mulligan LM, Kwok JBJ, Healey CS et al. Germline mutations of the ret proto-oncogene in multiple endocrine neoplasia type 2A.1993; 363:458-60.
16. Donis-Keller H, Dou S, Chi D et al. Mutations in the ret proto-oncogene are associated with MEN 2A and FMTC. Hum Mol Genet 1993; 2:851-6.
17. Hofstra RMW, Landsvater RM, Ceccherini I et al. A mutation in the ret proto-oncogene associated with multiple endocrine neoplasia type 2B and sporadic medullary thyroid carcinoma. Nature 1994; 367:375-6.
18. Franssila K. Value of histologic classification of thyroid cancer. Acta Pathol Microbiol Scand 1971; Suppl.225:5-76.
19. Hill CS, Ibanez ML, Samaan NA et al. Medullary (solid) carcinoma of the thyroid gland: an analysis of the M.D. Anderson Hospital Experience with patients with the tumor, its special features and its histogenesis. Medicine 1973; 52:141-71.
20. Steiner AL, Goodman AD, Powers SR. Study of a kindred with pheochromocytoma, medullary thyroid carcinoma, hyperparathyroidism, and Cushing's disease: MEN, type II. Medicine 1968; 47:371-409.
21. Gagel RF, Jackson CE, Ponder BAJ et al. Multiple Endocrine Neoplasia Type 2 Syndromes: Nomenclature Recommendations from the Workshop Organizing Committee. Henry Ford Hosp Med

J 1989; 37:99.
22. Farndon JR, Leight GS, Dilley WG et al. Familial medullary thyroid carcinoma without associated endocrinopathies: a distinct clinical entity. Br J Surg 1986; 73:278-81.
23. Nunziata V, Giannattasio R, di Giovanni G et al. Hereditary localized pruritis in affected members of a kindred with multiple endocrine neoplasia type 2A (Sipple's Syndrome). Clin Endocrinol 1989; 30:57.
24. Gagel RF, Levy ML, Donovan DT et al. Multiple endocrine neoplasia type 2a associated with cutaneous lichen amyloidosis. Ann Intern Med 1989; 111:802.
25. Mulligan LM, Eng C, Attie T et al. Diverse phenotypes associated with exon 10 mutations of the ret proto-oncogene. Hum Mol Gen 1994; 3:2163-67.
26. Ponder BAJ, Ponder MA, Coffey R et al. Risk estimation and screening in families of patients with medullary thyroid carcinoma. Lancet 1988; 1:397-400.
27. Wells SA Jr, Chi DD, Toshima K et al. Predictive DNA testing and prophylactic thyroidectomy in patients at risk for multiple endocrine neoplasia type 2A. Ann Surg 1994; 220:237-50.
28. Lips CJM, Landsvater RM, Hoppener JWM et al. Clinical screening as compared with DNA analysis in families with multiple endocrine neoplasia type 2A. N Engl J Med 1994; 331:828-35.
29. Baylin SB, Hsu SH, Gann DS. Inherited medullary thyroid carcinoma: a final monoclonal mutation in one of multiple clones of susceptible cells. Science 1978; 199:429-31.
30. Telander RL, Zimmerman D, van Heerden JA et al. Results of early thyroidectomy for medullary thyroid carcinoma in children with multiple endocrine neoplasia type 2. J Ped Surg 1986; 21:1190-9.
31. Carney JA, Sizemore GW, Hayles AB. Multiple endocrine neoplasia, type 2B. Pathobiol Ann 1978; 8:105-53.
32. LiVolsi VA. Surgical Pathology of the Thyroid. Philadelphia: WB Saunders, 1990.
33. Russell CF, Van Heerden JA, Sizemore GW et al. The surgical management of medullary thyroid carcinoma. Ann Surg 1983; 197:42-8.
34. Ellenhorn JDI, Shah JP, Brennan MF. Impact of therapeutic regional lymph node dissection for medullary carcinoma of the thyroid gland. Surgery 1993; 114:1078-81.
35. Dralle H, Scheumann GFW, Proye C et al. The value of lymph node dissection in hereditary medullary thyroid carcinoma: a retrospective, European multicentre study. J Int Med 1995; 238:357-61.
36. Melvin KEW, Tashjian AH Jr., Cassidy CE et al. Cushing's syndrome caused by ACTH- and calcitonin-secreting medullary carcinoma of the thyroid. Metabolism 1970; 19:831-8.
37. Chen H, Doppman JL, Chrousos GP et al. Adrenocorticotropic

hormone-secreting pheochromocytomas: the exception to the rule. Surgery 1995; 118:988-94.
38. Becker KL Nash D, Silva OL et al. Increased serum and urinary calcitonin levels in patients with pulmonary disease. Chest 1981; 79:211-6.
39. Simmons RE, Hjelle JT, Mahoney C et al. Renal metabolism of calcitonin. Am J Physiol 1988; 254:F593-600.
40. Wells SA, Dilley WG, Farndon JA et al. Early diagnosis and treatment of medullary thyroid carcinoma. Arch Intern Med 1985; 145:1248-52.
41. Amara SG, Jonas V, Rosenfeld MG et al. Alternative RNA processing in calcitonin gene expression generates mRNAs encoding different polypeptide products. Nature 1982; 298:240-4
42. Tache Y, Holzer P, Rosenfeld MG, eds. Calcitonin gene-related peptide: the first decade of a novel pleiotropic neuropeptide. Annals of the New York Academy of Sciences 1992; 657:561.
43. Mason RT, Shulkes A, Zajac JD et al. Basal and stimulated release of calcitonin gene- related peptide (CGRP) in patients with medullary thyroid carcinoma. Clin Endocrinol (Oxf) 1986; 25:675-85.
44. Schifter S. Calcitonin gene related peptide and calcitonin as tumour markers in MEN 2 family screening. Clin Endocrinol (Oxf) 1989; 30:263-70.
45. Roos BA, Huber B, Birnbaum RS et al. Medullary thyroid carcinomas secrete a non calcitonin peptide corresponding to the carboxyl-terminal region of pre-procalcitonin. J Clin Endocrinol Metab 1983; 56:802-7.
46. Modigliani E, Vasen HM, Raue K et al. Pheochromocytoma in multiple endocrine neoplasia type 2: European study. J Intern Med 1995; 238:363-7.
47. Lairmore TC, Ball DW, Baylin SB et al. The management of pheochromocytoma in patients with multiple endocrine neoplasia type 2 syndromes. Ann Surg. 1993; 217:595-601.
48. Mulligan LM, Marsh DJ, Robinson BG et al. Genotype-phenotype correlation in multiple endocrine neoplasia type 2: report of the international ret mutation consortium. J Intern Med 1995; 238:343-6.
49. Thomas JE, Rooke ED, Kvale WF. The neurologist's expereince with pheochromocytoma: a review of 46 cases. J Urol 1974; 111:715-21.
50. Hermann H, Mornex R. Human tumors secreting catecholamines. New York: Macmillan, 1964:1-14.
51. Neumann HP, Berger DP, Sigmund G et al. Pheochromocytomas, multiple endocrine neoplasia type 2, and von Hippel-Lindau disease New Engl Med 1993; 329:1531-8.
52. Keiser HR. Pheochromocytoma and other diseases of the sympathetic nervous system. In: Becker KL, ed. Principles and Practice

of Endocrinology and Metabolism. 1st ed. Philadelphia: JB Lippincott, 1990:676-682.
53. Glowniack JV, Shapiro B, Sisson, JC et al. Familial extra-adrenal pheochromocytoma: a new syndrome. Arch Intern Med 1985; 145: 257-61.
54. Crossey PA, Richards FM, Foster K et al. Identification of intragenic mutations in the Von Hippel-Lindau disease tumour suppressor gene and correlation with disease phenotype. Hum Mol Genet 1994; 3:1303-8.
55. Evans DB, Lee JE, Merrell RC et al. Adrenal medullary disease in multiple endocrine neoplasia type 2. Endocrinol Metab Clin N Am 1994; 23:167-76.
56. Hamilton BP, Landsberg L, Levine RJ. Measurement of urinary epinephrine inscreening for pheochromocytoma in multiple endocrine neoplasia 2. Am J Med 1978; 65:1027-32.
57. Feldman JM. Phenylethanloamine-N-methyltransferase activity determines the epinephrine concentration of pheochromocytomas. Res Commun Chem Pathol Pharmacol 1981; 34:389-98.
58. Kimura N, Miura Y, Nagatsu I et al. Catecholamine synthesizing enzymes in 70 cases of functioning and nonfunctioning pheochromocytoma and extra-adrenal paraganglioma. Virchows Arch A Pathol Anat Histopathol 1992; 421:25-32.
59. DeLellis RA, Wofe HJ, Gagel RF et al. Adrenal medullary hyperplasia. A morphometric analysis in patients with familial medullary thyroid carcinoma. Am J Pathol 1976; 83:177-96.
60. Melvin KEW, Tashjian AH Jr, Miller HH. Studies in familial (medullary) thyroid carcinoma. Rec Prog Horm Res 1972; 28:399-470.
61. Mulligan LM, Eng C, Healey CS et al. Specific mutations of the ret proto-oncogene are related to disease phenotype in MEN 2A and FMTC. Nat Genet 1994; 6:70-4.
62. Carney JA, Roth SI, Heath H III et al. The parathyroid glands in multiple endocrine neoplasia type 2B. Am J Pathol 1980; 99:387-98.
63. Gagel RF, Tashjian AH Jr., Cummings T et al. The clinical outcome of prospective screening for multiple endocrine neoplasia type 2a. An 18 year experience. N Engl J Med 1988; 318:478-84.
64. Eng C, Smith DP, Mulligan LM et al. A novel point mutation in the tyrosine kinase domain of the ret proto-oncogene in sporadic medullary thyroid carcinoma and in a family with FMTC. Oncogene 1995; 10:509-13.
65. Bugalho MJM, Cote GJ, Khoran S et al. Identification of a polymorphism in exon 11 of the ret proto-oncogene. Hum Mol Genet 1994; 3:2263.
66. Wohlik N, Cote GJ, Bughalho D et al. Relevance of ret proto-oncogene analysis in sporadic MTC. (Abstract) 77th Endocrine Society Meeting, 1995: 464.
67. Zedenius J, Wallin G, Hamberger B et al. Somatic and MEN 2A

de novo mutations identified in the ret proto-oncogene by screening of sporadic MTC's. Hum Mol Genet 1994; 3:1259-62.
68. Blaugrund JE, Johns MM, Ebyl YJ et al. ret proto-oncogene mutations in inherited and sporadic medullary thyroid cancer. Hum Mol Genet 1994; 3:1895-7.
69. van Heerden JA, Grant CS, Gharib H et al. Long term course of patients with persistent hypercalcitoninemia after apparent curative primary surgery for medullary thyroid carcinoma. Ann Surg 1990; 212:395-401.
70. Tisell LE, Hansson G, Jansson S et al. Reoperation in the treatment of asymptomatic metastasizing medullary thyroid carcinoma. Surgery 1986; 99:60-6.
71. Moley JF, Wells SA, Dilley WG et al. Reoperation for recurrent or persistent medullary thyroid cancer. Surgery 1993; 114:1090-5.
72. Saad MF, Ordonez NA, Rashid RK et al. Medullary carcinoma of the thyroid: a study of the clinical features and prognostic factors in 161 patients. Medicine 1984; 63:319-42.
73. Samaan NA, Schultz PN, Hickey RC. Medullary thyroid carcinoma: prognosis of familial versus sporadic disease and the role of radiotherapy. J Clin Endocrinol Metab 1988; 67:801-5.
74. Wu LT, Averbuch SD. Chemotherapy of Advanced Thyroid Cancer. In: Cobin RH, Srota DK, eds. Malignant Tumors of the Thyroid. New York: Springer-Verlag, 1992:204.
75. Wu L-T, Averbuch SD, Ball DW et al. Treatment of advanced medullary thyroid carcinoma with a combination of cyclophosphamide, vincristine and dacarbazine. Cancer 1994; 73:432-6.
76. Gagner M, Lacroix A, Prinz RA et al. Early experience with laparoscopic approach for adrenalectomy. Surgery 1993; 114:1120-4.
77. Bergholm U, Adami H-O, Auer G et al. Histopathologic characteristics and nuclear DNA content as prognostic factors in medullary thyroid carcinoma. Cancer 1989; 64:135-42.
78. Saad MF, Fritsche HA, Samaan NA. Diagnostic and prognostic values of carcinoembryonic antigen in medullary carcinoma of the thyroid. J Clin Endocrinol Metab 1984; 58:889-94.
79. Mendelsohn G, Wells SA, Baylin, SB. Relationship of tissue carcinoembryonic antigen and calcitonin to tumor virulence in medullary thyroid carcinoma. Cancer 1984; 54:657-62.
80. Ball DW, de Bustros A, Baylin SB. Medullary Thyroid Carcinoma. In: Braverman LE, Utiger RD, eds. The Thyroid. Philadelphia: Lippincott 1995:946-960.

CHAPTER 2

Mutations in Ret in MEN 2

B.A.J. Ponder and M.A. Pierotti

The different clinical varieties of MEN 2 (MEN 2A, MEN 2B, FMTC—see chapter 1) are associated with distinct types of mutation in the ret proto-oncogene. This genetic information coupled with a basic understanding of the ways in which the different types of mutation may affect the activity of the ret protein, has already provided interesting insights into the biology of receptor tyrosine kinases. A more detailed elucidation of the genotype: phenotype correlations in terms of different signaling pathways in different cell types provides a challenge for the future.

The germline mutations in the ret gene which predispose to MEN 2 and which have been identified to date, fall into three groups (Fig. 2.1): (i) mutations in the cysteine-rich region of the extracellular domain; (ii) mutations which affect the substrate binding pocket in Hanks domain VIII of the tyrosine kinase region; and (iii) mutations which affect other domains of the tyrosine kinase region. Each will be described in turn.

MUTATIONS IN THE CYSTEINE-RICH REGION OF THE EXTRACELLULAR DOMAIN

Description of the Mutations

The extracellular portion of the ret protein contains 27 cysteine residues, 26 of which are precisely conserved between man and mouse,[1] and also highly conserved in chicken[2] and Drosophila[3] ret. Mutations in MEN 2 have been described in 5 of the 6 cysteines closest to the cell membrane: cysteine codons 609, 611, 618, 620 and 634. The mutations are missense mutations, leading to

Genetic Mechanisms in Multiple Endocrine Neoplasia Type 2, edited by Barry D. Nelkin.

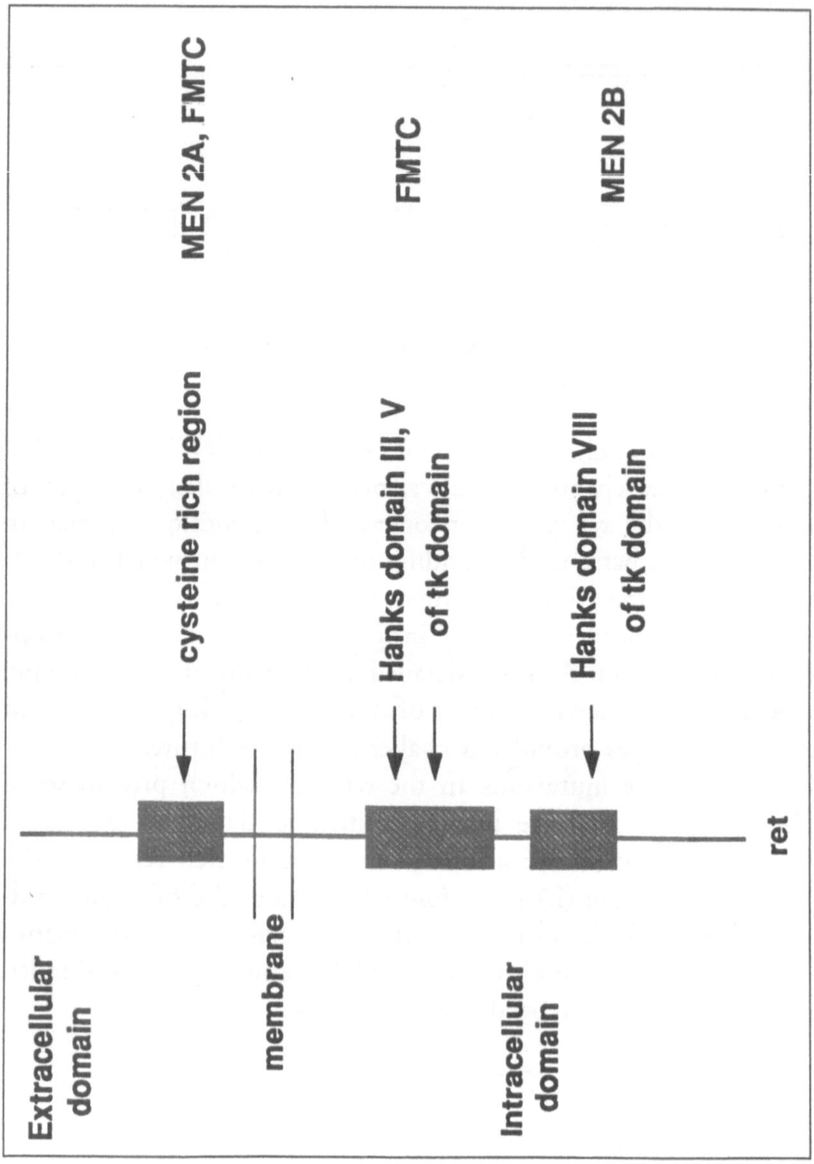

Fig. 2.1. Diagram of the ret protein and the sites of MEN 2 mutations.

the replacement of a single cysteine by another amino acid. Although there is also a cysteine at 630, involvement of this cysteine in germline mutations has not been reported.

The distribution of mutations in the different cysteine codons is shown in Figure 2.2. There is a clear phenotype-genotype correlation related to the cysteine codon that is mutated, which has been found independently by several groups.[4] Adopting the definitions of MEN 2A and FMTC set out in Table 2.1,[5] the great majority (87%) of mutations in MEN 2A families involve cys634, whereas the mutations in FMTC are more evenly distributed among the cysteines, with a majority in cysteines other than cys634. The distribution of mutations in a large numbers of families which, because of small size or missing clinical information, could not be confidently classified as either MEN 2A or FMTC, falls as expected between these two patterns.

There is also some evidence, as yet unconfirmed, which suggests a further phenotype: genotype correlation based on the particular amino acid which replaces the cysteine. At codon 634, the most frequent amino acid substitution (52% in one series)[4] is arginine. Comparison of families having a cys634 arg mutation with families in which there was a non-arg substitution at codon 634, showed a very strong correlation between the cys634 arg mutation and the presence of parathyroid involvement in at least one member of the family.[5] No such effect was seen for pheochromocytoma. Haplotyping of the families with cys634arg mutation and parathyroid disease[6] showed that the mutations were almost certainly of

Table 2.1. MEN2A and FMTC classification

MEN 2A	
	Families with MTC and either phaeo or parathyroid disease, or both.
MEN 2B	
	Families with MTC with or without phaeo and with characteristic clinical abnormalities but without parathyroid disease.
FMTC	
	Families with a minimum of 4 members with MTC and without evidence of phaeo and parathyroid disease in screening of living affected and at risk individuals.
Other	
	Families with fewer than 4 members with MTC but no individuals with phaeo or parathyroid disease on biochemical screening or families with any number of MTC without phaeo and parathyroid disease but in which screening for phaeo and/or parathyroid disease was incompletely documented.

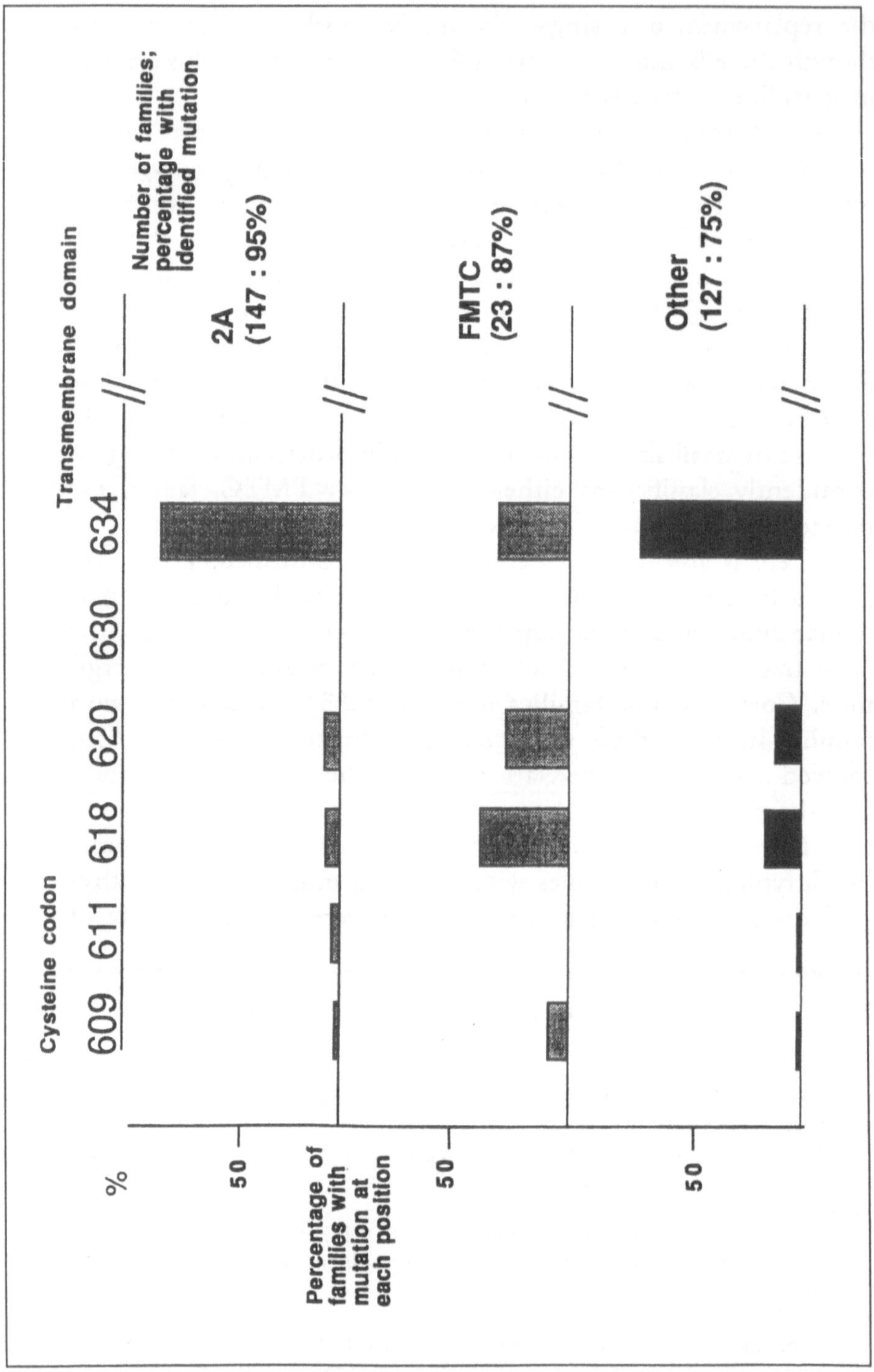

Fig. 2.2. Proportion of MEN 2A and FMTC mutations associated with different cysteine codons in ret (data from ref. 4).

independent origin, arguing against another polymorphism in linkage disequilibrium with the cys634arg mutation being responsible for this observed effect. Because this finding is unexpected, and it has not yet been confirmed, it should be treated with caution. Unfortunately, replication of the study is difficult because of the need for stringent phenotypic data from a large set of families.

The Same Mutations Are Described in MEN 2 and in Hirschsprung's Disease

Unexpectedly, cys609 mutations have also been reported in families with Hirschsprung's disease.[7,8] This is unexpected because the Hirschsprung's phenotype results from loss of function mutations in one allele of ret, whereas the MEN 2 mutations, as discussed below and in Chapter 3, are thought to be activating. Even more surprising, several families have been described in which MEN 2A or FMTC cosegregate with Hirschsprung's disease.[9,10] Several individuals in these families display both phenotypes. In each of 5 such families reported, there is a cys618 or cys620 mutation; careful analysis by sequencing of the rest of the ret coding sequence has revealed no other abnormality. The conclusion must be that in some individuals in some families, the same cys618 or cys620 mutation can result in both MEN 2 and Hirschsprung's disease phenotypes simultaneously.

The Extracellular Domain Cysteine Mutations Result in Activation of Ret

Experiments in which constructs containing proto-ret in the wild-type, MEN 2A (cys634) or MEN 2B (met918thr) mutant forms have been transfected into NIH 3T3 cells and PC12 cells provide evidence that the MEN 2A mutations lead to activation of the ret kinase. This evidence is of three types:

(1) Biological. Under conditions in which the wild-type construct yields no effect, both MEN 2A and MEN 2B mutant ret induce transformed foci in NIH 3T3 cells and differentiation of PC12 cells (see chapter 3). Santoro et al[11] tested three different codon 634 mutations: cys634arg, cys634tyr and cys634 trp, and the MEN 2B met918thr mutation, and found them to be of roughly equal effect. Other data[12] in which the transforming effect of the cys634arg and met918thr mutations have

been compared over a range of amounts of DNA transfected suggest, however, that the effect of the MEN 2B mutation in this system is consistently weaker than the MEN 2A.

(2) Steady-state tyrosine phosphorylation of the ret protein. MEN 2A and 2B, but not wild-type, ret protein from the NIH 3T3 transfectants is phosphorylated on tyrosine.

(3) Tyrosine kinase activity. Assay of ret protein in immunoprecipitates from transient transfection of HeLa cells shows that while the wild-type protein has some activity in auto kinase assays and in phosphorylation of the exogenous substrate myelin basic protein, the activity of the MEN 2A mutant protein is consistently higher.[12]

The Extracellular Domain Cysteine Mutations Activate Ret by Covalent Dimerization

The consistent involvement of specific cysteines in the mutations in MEN 2A and FMTC suggested a mechanism in which these cysteines would normally be paired in intramolecular disulfide bonds, so that the loss of one cysteine by mutation would result in the partner cysteine forming an intermolecular bond with the corresponding free cysteine on an adjacent ret molecule (Fig. 2.3). The resulting covalent dimerization would lead to constitutive activation of the ret tyrosine kinase domain, according to the model accepted for other receptor tyrosine kinases. This proposal predicts that in C-cell tumors containing a MEN 2A mutant ret, or in cells transfected with a ret cysteine mutant, ret dimers should be present which should be demonstrable by Western blotting of non-reducing gels. This prediction has been confirmed.[11]

THE MUTATION IN CODON 918, IN HANKS DOMAIN VIII OF THE TYROSINE KINASE REGION

This mutation is exclusively associated with MEN 2B.[13-15] ninety-three percent of MEN 2B families (many families consist of only one affected individual) reported so far have the identical mutation: met918thr.[4] The mutations in three families, in which the MEN 2B phenotype has been well-documented, are still unknown. In at least 2 of these families, careful analysis of the remainder of the ret gene has revealed no abnormality.

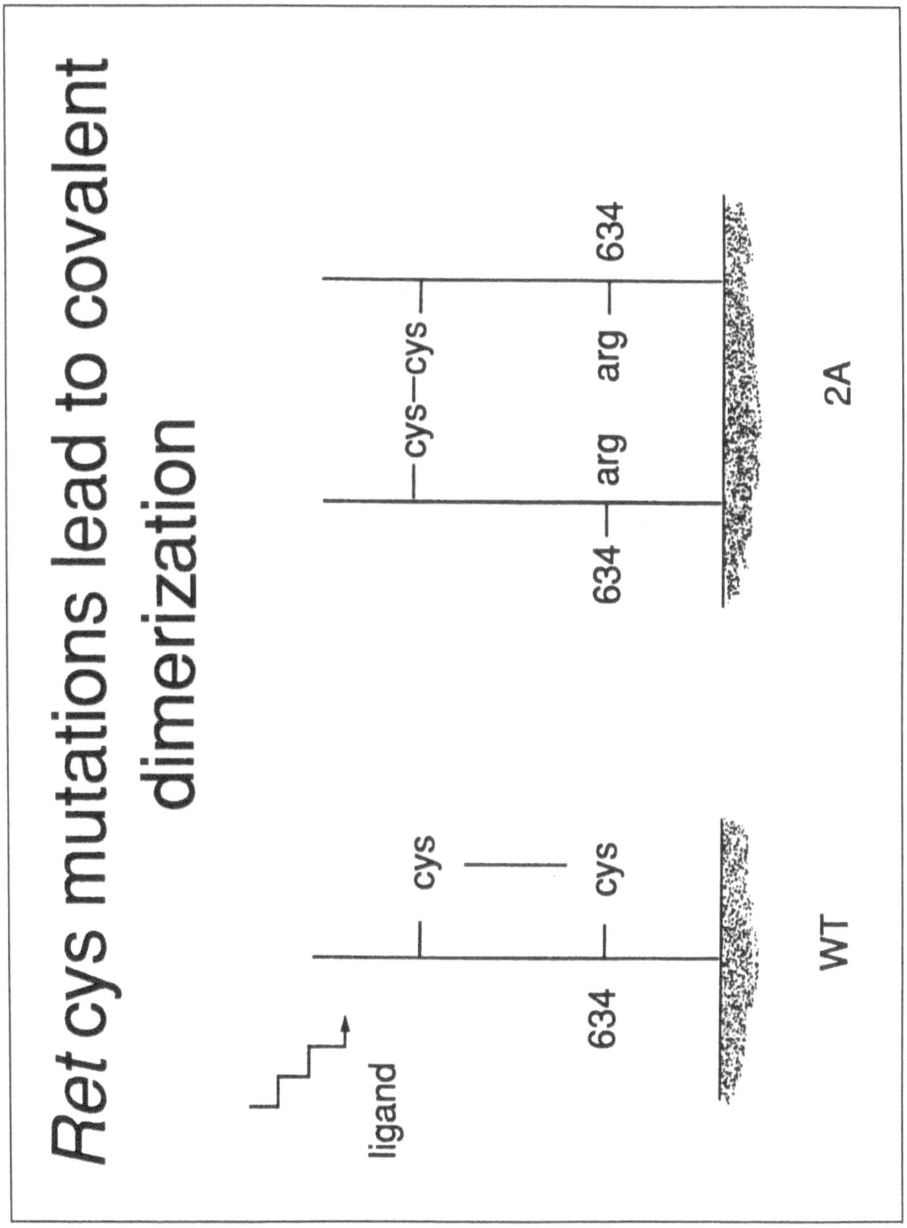

Fig. 2.3. Intermolecular disulfide bridges in activation of ret by MEN 2A mutations.

Residue 918 of ret is predicted from modeling studies to lie at the base of a pocket in the protein, which is involved in substrate binding.[13] The substitution of threonine for methionine would alter the dimensions of the pocket, and hence the substrate specificity. Furthermore, almost all receptor tyrosine kinases have methionine at position 918 (as does ret), whereas almost all cytoplasmic tyrosine kinases have threonine. This leads to two predictions: (1) receptor and cytoplasmic tyrosine kinases will differ in their substrate specificity; and (2) the met918thr mutation in MEN 2B will alter the specificity of ret away from that of a receptor tyrosine kinase, towards that of a cytoplasmic tyrosine kinase. There is now experimental evidence to support both predictions.

Using degenerate peptide libraries, Cantley's group[16] showed that the preferred peptide substrate for receptor tyrosine kinases has hydrophobic amino acids at both the +1 and +3 positions downstream of the target tyrosine, whereas the cytoplasmic tyrosine kinases prefer a hydrophilic amino acid at +1 and hydrophobic at +3. These different amino acid contexts flanking the tyrosine provide, in turn, preferred substrates for different groups of SH2 domains and hence, the possibility of different pathways of downstream signaling. When wild-type and MEN 2B mutant Ret proteins were compared for their ability to phosphorylate model substrates for a receptor tyrosine kinase (EGFR) and two different cytoplasmic tyrosine kinases (abl and src), the MEN 2B mutant protein showed a clear shift in specificity towards the substrates preferred by the cytoplasmic kinases.[16] The inference, that the MEN 2B mutation has shifted the pathway of downstream signaling from the ret kinase, is supported by evidence that activated MEN 2B ret differs from wild-type both in the pattern of tyrosine phosphorylation of the ret protein itself and in the pattern of tyrosine phosphoproteins (presumably involved in downstream signaling), which is seen in cell extracts.[11]

Does the met918thr Mutation Have Effects Other than on Substrate Specificity?

It is an unresolved question whether the effects of the met918thr mutation are restricted to altering the substrate specificity of the ret kinase, and whether this is therefore the sole explanation for the MEN 2B phenotype. The results of transfection experiments in NIH 3T3 cells suggest that the catalytic activity of the tyrosine kinase domain may also be enhanced, but the interpretation of these experiments is uncertain. When proto-ret constructs encoding either a wild-type, MEN 2A or 2B mutant protein are transfected into NIH 3T3 cells, under conditions in which the wild-type ret gives no transformed foci, the MEN 2B mutant ret induces focus formation, at a level variously reported to be comparable[11] or less than[12] that obtained with the MEN 2A mutant. As was described above, the MEN 2B ret protein in these experiments is also active by the criteria of phosphorylation on tyrosine and tyrosine kinase activity towards other proteins. NIH 3T3 cells are not thought to secrete ret ligand, and Santoro et al[11] reported that, even with the use of cross-linking agents, they was unable to demonstrate the presence of dimers of MEN 2B mutant protein in the transfected cells. They therefore concluded that the MEN 2B mutation causes activation of the kinase through an intramolecular mechanism. Another approach is to use ret/PTC constructs to compare the activity of the wild-type and 2B mutant tyrosine kinase domains under conditions in which a consistent degree of activation is presumably provided by the PTC dimerization. Ret/PTC is a rearranged form of ret in which the 5' end of the gene encoding the extracellular and transmembrane domains is replaced by one of several sequences which have in common that they encode dimerization domains. The resulting protein is cytoplasmic, and consists of the tyrosine kinase domain constitutively activated by the dimerization). In these experiments,[12] the MEN 2B mutant tyrosine kinase shows a higher level of steady state tyrosine phosphorylation, a higher tyrosine kinase activity against exogenous substrate and a slightly higher transforming activity in NIH 3T3 cells than the corresponding wild-type form. If correct, this suggests that the met918thr mutation does indeed alter the properties of the Ret tyrosine kinase domain in ways other than simply specificity of substrate binding. It remains unclear whether MEN 2B proto-Ret has activity in vivo independent of ligand-induced dimerization; and if so, whether and to what extent a further increase in activity might follow ligand binding. Resolution of this is likely to be important for understanding the mechanism by which the MEN 2A and MEN 2B mutations are involved in tumorigenesis, as discussed below.

MUTATIONS ELSEWHERE IN THE TYROSINE KINASE DOMAIN

At the time of writing, only two other germline mutations have been reported to predispose to MEN 2. Each is uncommon and so far associated exclusively with FMTC. They are glu768asp, in Hanks domain III,[17] and val804leu, in Hanks domain V.[18]

The glu768asp mutation involves a glu residue which is highly conserved among different receptor tyrosine kinases in man, and across species. Modeling based on the structure of cyclic AMP dependent kinase suggests that glu768 lies in a region of an alpha-helical turn which stabilizes the highly conserved glu and lys residues which are involved in ATP binding.[17] The predicted effect of the mutation is to activate the kinase: our own preliminary data from transfection experiments are consistent with this (Smith et al, unpublished). Modeling also suggests a possible effect on substrate specificity: this is being tested experimentally using the peptide library screening described earlier in relation to MEN 2B. Since the great majority of families with FMTC result from activating mutations in the extracellular cysteines, one might speculate that a simple activating effect would be sufficient to explain the glu768asp mutation; but the occurrence of this mutation in sporadic tumors also raises, equally speculative, the question of altered substrate specificity (see below).

The val804leu mutation affects a residue which is conserved among species, but no modeling of its possible effects has been reported.

OCCURRENCE OF THE SAME MUTATIONS IN SPORADIC MTC AND PHEOCHROMOCYTOMA

All inherited cancers have a sporadic, non-hereditary, counterpart. Generally, the same mutations are seen in the sporadic tumors, as somatic mutations, as are seen in the germline in familial cases. In MEN 2, this seems to be only partly true. The extracellular cysteine mutations characteristic of MEN 2A and FMTC have been found as somatic mutations (that is, present in tumor but not blood of the same individual) only rarely in sporadic MTC and sporadic pheochromocytomas.[19-23] By contrast, the met918thr mutation characteristic of MEN 2B is found in 25% or more of sporadic MTC, and about 10% of sporadic pheochromocytomas. The glu768asp mutation is intermediate in frequency.[19-23]

The difference in frequency of MEN 2A and 2B-type mutations in sporadic tumors suggests the following possible model for the role of these mutations in tumor formation in MEN 2.

A Model of Tumorigenesis in MEN 2 and Related Sporadic Tumors

The extracellular cysteine mutations

The inappropriate (ligand independent) activation of ret in embryogenesis may perturb the orderly development of the neuroectodermal lineage from which the C-cells and adrenal medulla are derived. As a result, these cells may fail to shut off division as they reach their final position, giving rise to the hyperplasia from which the tumors eventually develop. The absence of developmental abnormalities in association with these mutations (in contrast to MEN 2B) may be because, although the timing of activation of ret is inappropriate, the downstream signaling pathways are normal. Somatic mutations in the cysteine codons may rarely give rise to tumors because the mature C-cells and adrenal medullary cells are no longer susceptible to the activation of ret. Possibly, ret is already active in these cells because of the presence of ligand in the mature tissue; or possibly, at this developmental stage, the activation of ret has different consequences. This latter explanation is consistent with experimental observations that transfection of proto-ret into PC12 cells (derived from adrenal medulla) induced partial differentiation; and that in some cases induction of differentiation in cultured C-cell tumor lines can be accompanied by an increase in ret expression (see chapter 8).[24]

The met918thr mutation of MEN 2B

The interpretation of the consequences of this mutation is less clear. The mutation alters the substrate specificity of the ret tyrosine kinase; and this is the most plausible explanation for the spectrum of developmental abnormalities that is seen. However, there are some experimental results, described above, which, although far from conclusive, suggest that the mutation may also confer a degree of ligand independent activation and, possibly more relevant, an increased catalytic activity of the tyrosine kinase domain. Either of these could be significant in the effects of both the germline and somatic met918thr mutations, and increased rather than altered activity might account for the frequency of met 918thr mutations in sporadic tumors.

The absence of parathyroid involvement in MEN 2B is a further puzzle. One explanation could be lack of the ret ligand in parathyroid tissue (so that MEN 2A mutations would be effective, but MEN 2B would not, assuming they are ligand dependent). Alternatively, if the altered signaling pathways are important for MEN 2B tumorigenesis, these pathways may be absent in parathyroid.

Clinical Implications

MEN 2A shows considerable variation both between and within families in the age at onset of tumors and the spectrum of tissues (thyroid C cells, adrenal medulla, parathyroid) involved (see chapter 1). Prediction of the phenotype for a given individual might be helpful in determining screening and management. The genotype: phenotype correlations which have been noted in MEN 2A and FMTC families are, however, not sufficiently consistent to be useful for clinical prognosis.

The MEN 2B phenotype, while usually distinctive to an experienced observer, is sometimes questioned in a child or young adult with an unusual face, thickened corneal nerve fibers, or chronic bowel motility disturbance, even in the absence of thyroid tumor. In the authors' experience, one 12-year-old girl was submitted to thyroidectomy on the basis of a presumed MEN 2B phenotype, even though stimulated calcitonin testing was normal. Since the met918thr mutation has been found in 93% of MEN 2B cases examined so far, absence of this mutation should give pause for thought in the assessment of a possible MEN 2B phenotype. However, at least 3 individuals (2 from multiple case families) with well-documented MEN 2B have been reported to lack this mutation, so its absence provides probable rather than certain evidence against MEN 2B. It also remains quite possible that there are 'incomplete' or variant forms of the MEN 2B phenotype which are due to other mutations, but the risk of thyroid tumor in these putative cases would of course be uncertain.

A common clinical problem is the patient presenting with apparently sporadic MTC, who might yet possibly have hereditable disease, with implications for his or her family. Even though the chance of heritable disease may be quite low, the consequence of missing a chance of early diagnosis and surgical cure are such that many of these families are currently offered repeated biochemical screening. Now that mutation testing of the proband is possible, the question arises as to how certainly the absence of a mutation can exclude heritable disease, and thus the need to involve the family in screening. Since approximately 95% of families with the classical MEN 2A phenotype have mutations in codons 10 or 11 of ret, the lack of a family history coupled with lack of detectable mutation in these exons rules out MEN 2A with high probability, though not with certainty. Familial MTC (FMTC) is more difficult: mutations have been found in families with FMTC in exons 10, 11, 13 and 14, so the extent of the gene to be screened is greater; and a higher proportion of families with FMTC—in particular those with only 2 or 3 cases—have not yet been found to have ret mutations.[4] As a result, the lack of a detectable ret mutation provides for a less confident exclusion of familial disease in FMTC. Against this, it seems that the prognosis for clinically detected disease in these small MTC-only families may be better than in MEN 2A, so that the risks of omitting biochemical screening of family members may be less. The possible benefits of screening must always be weighed against the possible disadvantages that family members will become 'medicalized.'

As the spectrum of mutations in ret—and possibly in other genes—in MEN 2A and familial MTC becomes clearer, and the risk of disease associated with each mutation is better established, it will be possible to provide more firmly based advice about mutation screening in the apparently sporadic case. At the moment, it is probably justified to screen the exons in which mutations are known to occur—currently exons 10, 11, 13 and 14—in any sporadic case of MTC who has family members at risk. The decision to institute biochemical screening or not, should no mutation be found, is a matter for the clinician responsible for each family.

REFERENCES

1. Iwamoto T, Taniguichi M, Asai N et al. cDNA cloning of mouse ret proto-oncogene and its sequence similarity to the cadherin superfamily. Oncogene 1993; 8:1087-91.
2. Schuchardt A, Srinivas S, Pachnis V, Constantini F. Isolation and characterization of a chicken homolog of the c-ret proto-oncogene. Oncogene 1994; 10:641-49.
3. Sugaya R, Ishimara S, Hosoya T. A Drosophila homology of hu-

man proto-oncogene ret transiently expressed in embyronic neuronal precursor cells including neuroblasts and CNS cells. Mechanisms of Development 1994; 45:139-45.
4. Mulligan LM, Marsh DJ, Robinson BG et al. Genotype-phenotype correlation in MEN 2: Report of the International RET Mutation Consortium. J Intern Med 1995; 238:343-6.
5. Mulligan LM, Eng C, Healey CS et al. Specific mutations of the ret proto-oncogene are related to disease phenotype in MEN 2A and FMTC. Nat Genet 1994; 6:70-4.
6. Gardner E, Mulligan LM, Eng C et al. Haplotype analysis of MEN 2 mutations. Hum Mol Genet 1994; 3:1771-4.
7. Angrist M, Bolk S, Thiel B et al. Mutation analysis of the RET receptor tyrosine kinase in Hirschsprung's disease. Hum Mol Genet 1995; 4:821-30.
8. Attie T, Pelet A, Edery P et al. Diversity of RET proto-oncogene mutations in familial and sporadic Hirschsprung disease. Hum Mol Genet 1995; 4:1381-6.
9. Mulligan LM, Eng C, Attie T et al. Diverse phenotypes associated with exon 10 mutations in the RET proto-oncogene. Hum Mol Genet 1994; 3:2163-7.
10. Borst MJ, Van Camp JM, Peacock ML, Decker RA. Mutation analysis of multiple endocrine neoplasia type 2A associated with Hirschsprung's disease. Surgery 1995; 117:386-9.
11. Santoro M, Carlomagno F, Romano A et al. Activation of RET as a dominant transforming gene by germline mutations of MEN 2A and MEN 2B. Science 1995; 267: 381-3.
12. Borello MG, Smith DP, Pasini B et al. RET activation by germline MEN 2A and MEN 2B mutations. Oncogene 1995; 11:2412-27.
13. Hofstra RMW, Landsvater RM, Ceccherini I et al. A mutation in the ret proto-oncogene associated with multiple endocrine neoplasia type 2B and sporadic medullary thyroid carcinoma. Nature 1994; 367:375-76.
14. Carlson KM, Dou S, Chi D et al. Single missense mutation in the tyrosine kinase catalytic domain of the ret proto-oncogene is associated with multiple endocrine neoplasia type 2B. Proc Natl Acad Sci USA 1994; 91:1579-83.
15. Eng C, Smith DP, Mulligan LM et al. Point mutation within the tyrosine kinase domain of the ret proto-oncogene in multiple endocrine neoplasia type 2B and related sporadic tumours. Hum Mol Genet 1994; 3:237-41.
16. Songyang Z, Carraway III KL, Eck MJ et al. Catalytic specificity of protein tyrosine kinases is critical for selective signalling. Nature 1995; 373:536-9.
17. Eng C, Smith DP, Mulligan LM et al. A novel point mutation in the tyrosine kinase domain of the RET proto-oncogene in sporadic medullary thyroid carcinoma and in a family with FMTC.

Oncogene 1995; 10:509-13.
18. Bolino A, Schuffenecker I, Luo Y et al. RET mutations in exons 13 and 14 of FMTC patients. Oncogene 1995; 10:2415-9.
19. Eng C, Mulligan LM, Smith DP et al. Mutations of the RET proto-oncogene in sporadic medullary thyroid carcinomas. Genes Chromosom Cancer 1995; 12:209-12.
20. Zedenius J, Wallin G, Hamberger B et al. Somatic and MEN 2A de novo mutations identified in the RET proto-oncogene by the screening of sporadic MTCs. Hum Mol Genet 1994; 3:1259-62.
21. Blaugrund JE, Johns MMJ, Eby JY et al. RET proto-oncogene mutations in inherited and sporadic medullary thyroid cancer. Hum Mol Genet 1995; 3:1895-7.
22. Lindor NM, Honchel R, Khosla S, Thibodeau SN. Mutations in the RET proto-oncogene in sporadic pheochromocytomas. J Clin Endocrinol Metab 1995; 80:627-9.
23. Beldjord C, Desclaux-Arramond F, Raffin-Sanson M et al. The RET proto-oncogene in sporadic pheochromocytoma. Am J Hum Genet 1994; 55(suppl):A51.
24. Santoro M, Rosati R, Grieco M et al. The ret proto-oncogene is consistently expressed in human pheochromocytomas and thyroid medullary carcinomas. Oncogene 1990; 5:1595-8.

CHAPTER 3

Intracellular Signaling by the ret Tyrosine Kinase

Alfredo Fusco, Giancarlo Vecchio, Nina A. Dathan, Francesca Carlomagno, Pier Paolo Di Fiore and Massimo Santoro

BACKGROUND

Signaling by receptors with tyrosine kinase activity (RTK) plays an important role in the control of such cellular processes as cell growth, differentiation and motility. The binding of growth factors to RTKs promotes the activation of their intrinsic tyrosine kinase function and their interaction with a repertoire of intracellular molecules that elicit the appropriate biological response.[1] In some cases "gain of function" mutations lead to a constitutive activation of the receptor and, as a consequence, to a chronic stimulation of its intracellular signaling pathway.[2] Indeed, many members of the RTK gene superfamily were initially isolated as oncogenes that arose from mutations deregulating their kinase activity. Ret, a member of the RTK family, was first isolated as a transforming gene created by a recombination with the *rfp* gene during transfection of a T cell lymphoma DNA.[3] It is characterized by a tyrosine kinase domain, divided in two subdomains by a 27 amino acid long insertion, and by an extracellular domain which shows a 110 amino acid sequence resembling a repeat found in cadherins (Fig. 3.1).[4] At least two isoforms of ret, generated by alternative splicing, have been characterized.[4] The cloning of

Genetic Mechanisms in Multiple Endocrine Neoplasia Type 2A, edited by Barry D. Nelkin.

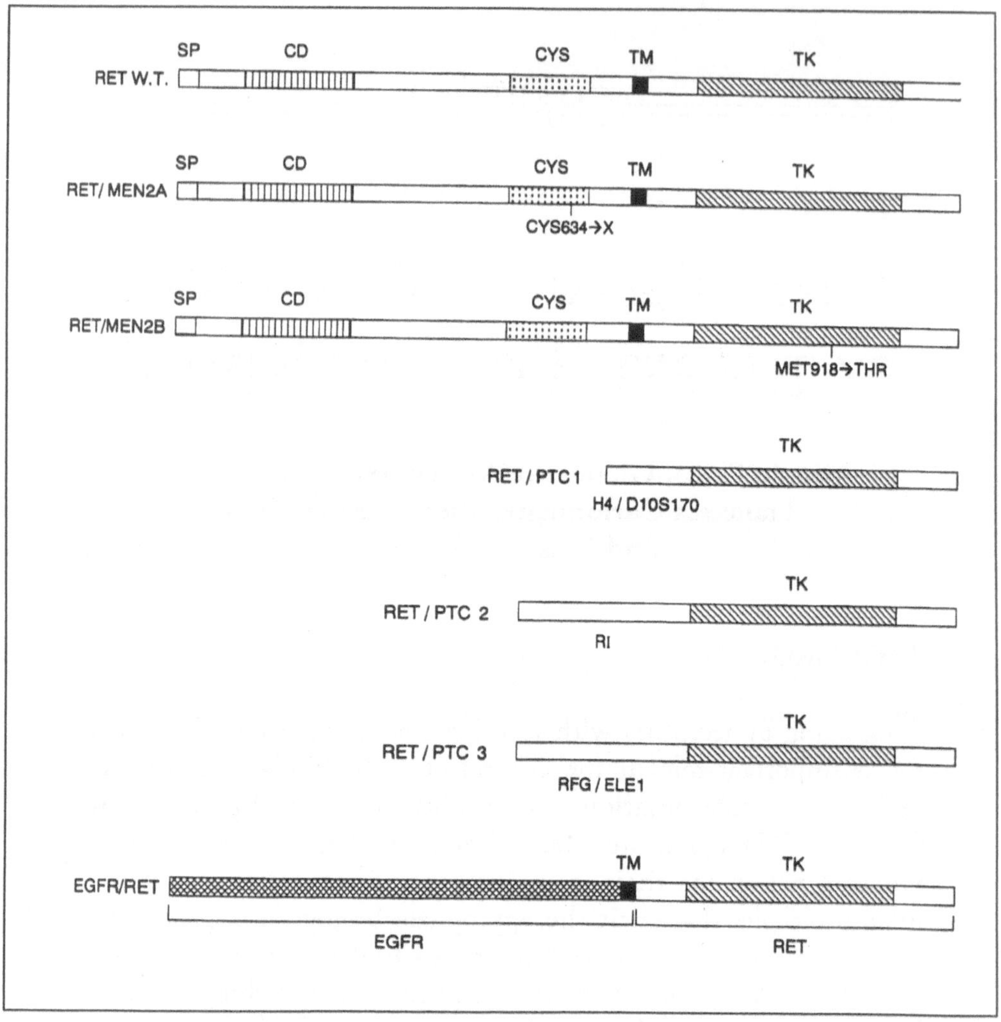

Fig. 3.1. Schematic representation of activated ret isoforms. Ret encodes for a transmembrane receptor characterized by a signal peptide (SP), a cadherin-like (CD) and a cysteine-rich (CYS) extracellular domain, a transmembrane domain (TM) and a cytoplasmic tyrosine-kinase domain (TK). Substitution of one of five cysteine residues located in the juxtamembrane region of ret and a Met to Thr substitution in the ret TK domain are the point mutations responsible for the inheritance of MEN 2A and MEN 2B syndromes. Ret activated isoforms in papillary thyroid carcinomas consist of chimeric oncogenes created by the fusion of its TK domain to the 5' portion of different activating genes: H4/D10S170, RI subunit of PKA, and RFG/ELE1 in the case of RET/PTC1, 2 and 3, respectively. An inducible form of RET (EGFR-RET) has been obtained by fusing the ret TK domain to the extracellular and transmembrane domain of the EGFR.

human,[4] murine,[5] chicken[6] and even *Drosophila melanogaster*[7] homologues demonstrated that ret is highly conserved during evolution.

The primary response to RTK-mediated signaling in most cells is the stimulation of cell proliferation. However, effects can differ according to the cell type. In this conceptual framework, the association of ret expression with the development of specific cell lineages, together with the observation that activation of ret in human tumors is limited to thyroid carcinomas and neural crest tumors, suggests a peculiar responsiveness of organ-specific mitogenic and/or differentiating pathways to the action of the ret kinase. The elucidation of these issues requires an understanding of the molecular events involved in the transduction of ret signals.

Expression of the ret Tyrosine Kinase Receptor

Expression studies suggest that the physiological function of ret is related to the differentiation of particular components of the nervous and excretory systems. During murine embryogenesis, ret expression is found in migrating neural crest cells, in the deriving dorsal root and sympathetic ganglia and in developing cranial ganglia. In the central nervous system, ret is present in the gangliar layer of the retina and in the ventral portion of the neural tube, where motor neurons are located.[8,9] In addition to the nervous system, ret expression is detected in the developing excretory system: specifically in the nephric duct, the ureteric bud and the collecting ducts of the kidney.[8,9] In several of these regions, ret expression is punctate in nature, which indicates that only certain subsets of cells contain ret mRNA.[8] A similar distribution of ret was observed during rat[10] and chicken[6] development. In adult animals, ret expression in neural structures is generally preserved but its expression in kidney is lost.[8,9] High levels of ret have been found in mouse and rat brain and salivary glands.[8,10] In adult rat brain, ret proteins are specifically detected in cerebellum and hypothalamus, and to a lesser extent, in the striatum and in the pons Varolii (de Franciscis, manuscript submitted). In addition, epithelial cells of thymus, follicular dendritic cells of the spleen and lymph nodes, some adrenal chromaffin cells and thyroid C-cells express ret in rats.[10] Also in humans, C-cells of the thyroid and thymus epithelial cells express the ret gene.[11,12]

Evidence is accumulating that ret signaling is critical in development of the structures where it is expressed. Sites of ret expression are evolutionary conserved and probably functionally important. Even the expression profile of the *Drosophila* homolog (D-ret) is closely related to the mammalian ret, being transiently expressed in neuroblasts and ganglial cells; in addition, its expression is altered in neurogenic mutants.[7] Targeted disruption of ret causes a lack of enteric neurons and severe defects in kidney development in homozygous mice.[13] Mutations that totally disrupt or partially change the structure of the tyrosine kinase domain of the ret protein have been described in patients affected by Hirschsprung's disease, a genetic disorder of neural crest development, characterized by the absence of intramural ganglion cells in the hindgut.[14,15] Finally, ret expression is associated with the differentiation of neuronal cells. A marked accumulation of ret protein was observed in motoneuron cell bodies during hypoglossal nerve regeneration (de Franciscis, manuscript submitted). Increased expression of ret is observed when the TT medullary carcinoma cell line or several neuroblastoma cell lines are induced to differentiate by cAMP or retinoic acid, respectively.[16,17] Taken together, the foregoing results leave little doubt that ret signaling is involved in the differentiation pathways of neural cells.

Ret Activation in Human Tumors

Ret is altered in two types of human neoplasias. Ret rearrangements, occuring in papillary thyroid carcinomas, and ret point mutations are associated with tumors characterizing the multiple endocrine neoplasia type 2A and 2B syndromes and familial medullary thyroid carcinomas (see chapter 2). The structure of the normal ret gene and all of its activated forms is shown in Figure 3.1. All of these activated forms have been shown to act as dominant oncogenes in NIH 3T3 cells.[18,19]

Ret is activated in about 20% of thyroid papillary carcinomas.[20-23] The fusion of the ret tyrosine kinase domain with the 5' terminal region of another gene, H4/D10S170,[24] yielded in the first case studied a chimeric gene designated ret/PTC1.[22] Both H4 and ret map on the long arm of chromosome 10 and their fusion is caused by a chromosomal inversion.[25] Recently, two other ret/PTC isoforms, designated ret/PTC2 and ret/PTC3, have been isolated from human thyroid papillary carcinomas. The 5' portion of the ret/PTC2 oncogene is represented by the regulatory subunit RI of the cAMP-dependent protein kinase A,[26] while a previously unknown gene, named RFG/ELE1, is fused to ret in the ret/PTC3 oncogene.[27,28] In each case, the genes fused to ret are ubiquitously expressed and, therefore, able to drive the expression of truncated forms of ret in thyroid follicular cells, which normally do not express this gene.[22] In addition, all the resulting oncogenic fusion proteins display a constitutive activation of their tyrosine kinase function; this may be a consequence of the deletion of the extracellular domain and of the presence, in the activating genes H4/D10S170 and RI, of coiled-coil structures, which could confer an ability to form dimers to the oncoproteins.[24,25] Furthermore, these ret oncoproteins are localized in the cytoplasm and therefore their associated enzymatic activity is translocated from the membrane to a novel cell compartment.[29] Ret rearrangements are restricted to thyroid carcinomas of the papillary histotype.[23] More than 100 non-papillary thyroid tumors and more than 500 non-thyroid neoplasias were shown to be negative for ret/PTC activation.[30] This suggests either that ret activation occurs only in thyroid cells, or that this event also occurs in other cells, but is unable to drive cells of non-thyroid origin to the neoplastic phenotype. A peculiar responsiveness of thyroid mitogenic pathways to the action of ret kinase could be responsible for its role as an oncogene in vivo. However, transgenic mice models showed that activated ret can induce several non-thyroid tumors.[31,32] Mammary adenocarcinomas have also been found in transgenic mice carrying the ret/PTC1 construct (Portella, manuscript in preparation).

To define the role of ret/PTC in the process of thyroid carcinogenesis, we infected a differentiated rat thyroid cell line, PC Cl 3, with a retrovirus carrying the ret/PTC1 oncogene. PC Cl 3 expresses the typical markers of thyroid differentiation (i.e., thyroglobulin synthesis and secretion, ability to trap iodide and dependency for growth on thyrotropin-TSH).[33] Upon infection with the ret/PTC1 virus, the PC Cl 3 cells became independent of TSH for growth and completely lost their differentiated functions. However, they did not acquire the malignant phenotype.[34] Similar effects were obtained when we infected PC Cl 3 cells with retroviruses carrying viral ras genes.[33] Introduction of both ret/PTC1 and v-ras oncogenes into the PC Cl 3 cell line resulted in the appearance of a malignant phenotype, i.e., the cells became capable of growing very efficiently in semisolid medium and were able to induce subcutaneous tumors in athymic mice within a very short latency period.[34]

RET SIGNALING PROPERTIES

Extracellular mitogenic and differentiating signals are intracellularly transduced through RTKs. Upon ligand binding, RTKs dimerize and their increased kinase activity leads to tyrosine phosphorylation of a variety of intracellular substrates. RTKs specifically interact with intracellular molecules (Fig. 3.2). Upon activation, receptors become autophosphorylated on tyrosine residues. These phosphorylated tyrosines can serve as docking sites for molecules containing src homology region 2 (SH2) domains, which can interact with the phosphotyrosine and surrounding residues of the receptor.[35] In some cases, these molecules, which may be enzymes, cytoskeletal proteins, etc., are subsequently phosphorylated by the activated receptor; this phosphorylation can modify their function. In addition, binding of SH2-containing enzymes to RTKs may be important in the targeting of these enzymes, since the substrates for these enzymes, such as in the case of PI3-K, PLC-γ and p120GAP, are often localized close to the plasma membrane (see below). In other cases, proteins with SH2 domains are devoid of intrinsic enzymatic activity, but upon interaction with the receptor, they function as "adaptors" for the recruitment of enzymes to the proximity of the plasma membrane.[1,35] These protein-protein interactions often occur through another domain, Scr homology region 3 (SH3), which interacts with proline-rich regions on proteins.[35]

The ret signaling pathway has been studied in NIH 3T3 cells, a well characterized cell system used for the study of different RTK pathways. Since a ligand for ret has not yet been identified, the strategy used was based on the generation of a chimeric receptor possessing the extracellular domain of the epidermal growth factor (EGF) receptor (EGFR) and the intracellular domain of ret (EGFR/ret chimera) (Fig. 3.1). With this chimera, we were able to examine the biological effects and biochemical activities of the ret kinase under controlled conditions of activation. We found that EGFR/ret was correctly synthesized and transported to the cell surface, where it was shown to be capable of binding EGF and transducing an EGF-dependent signal intracellularly.[36] The

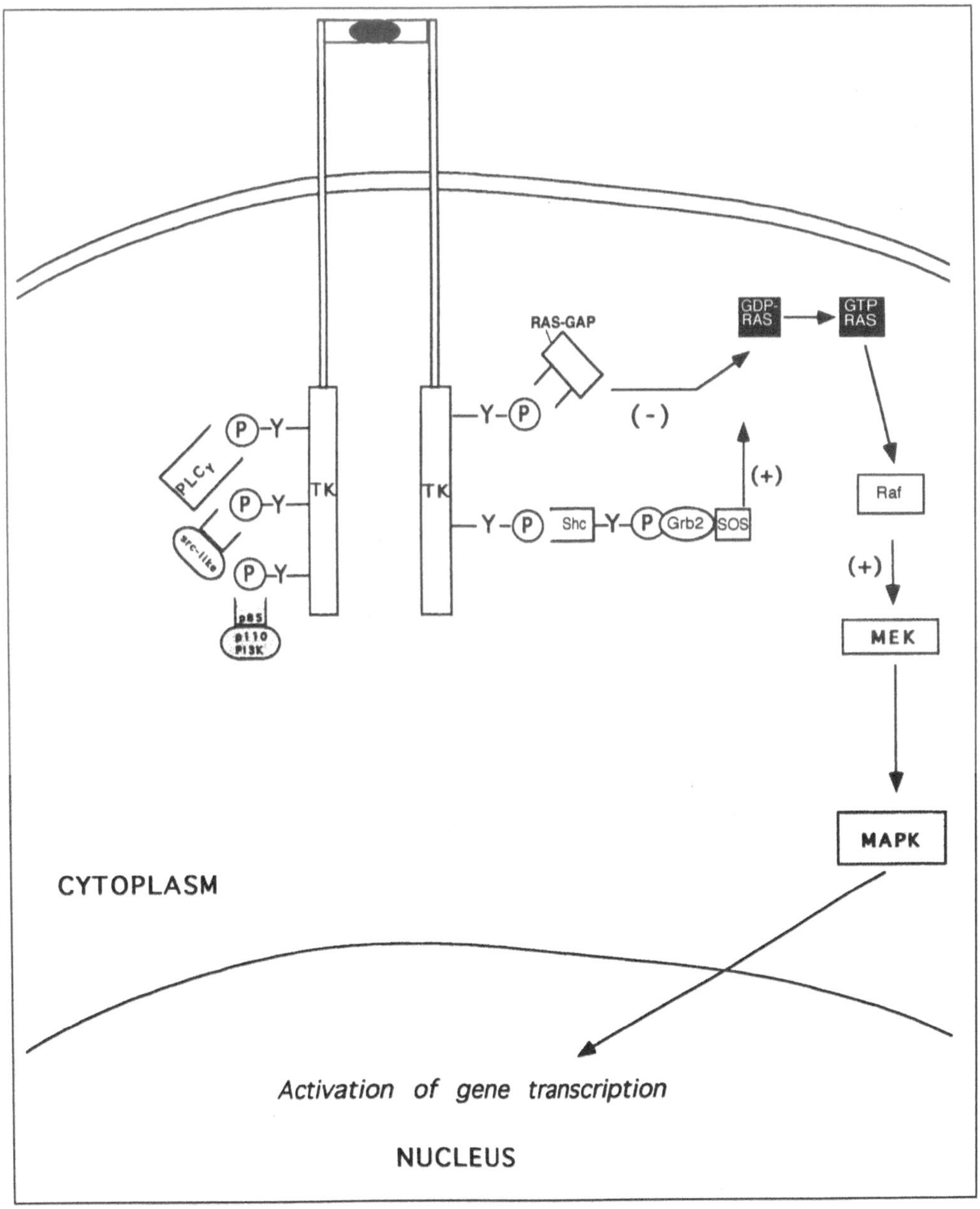

Fig. 3.2. Recruitment of intracellular transducers to activated tyrosine kinase receptors. Receptor-type tyrosine kinases undergo ligand induced dimerization, resulting in cross-phosphorylation on tyrosine residues. The tyrosine phosphorylated regions of the receptor provide specific binding sites for cytosolic proteins containing SH2 domains. These proteins can be endowed with enzymatic activities required for the transmission of signals to downstream effectors. Recruitment without phosphorylation is important, since the substrates for these enzymes are often localized close to the plasma membrane, such as in the case of PI3-K, PLC-γ and p120GAP. In addition, tyrosine phosphorylation by the receptor can modulate the activity of these enzymes.

transforming efficiency of this construct was very low in the absence of EGF, whereas upon the addition of EGF to the medium, it increased to about 10^3 FFU/pmol of DNA, a value comparable to that of most potent oncogenes (including ras and ErbB2).[36]

We investigated the ability of the EGFR/ret chimera to recruit several different pathways involved in the transduction of mitogenic signals in NIH 3T3 cells, by stimulating the cells either with EGF or PDGF and comparing the activity of the two receptors (see Table 3.1).

RET ACTIVATES PHOSPHOLIPASE C-γ BUT NOT PHOSPHATIDYLINOSITOL 3-KINASE (PTDINS-3K)

Effect of ret activation on PLCγ

RTKs stimulate the enzymatic activity of PLCγ (the γ isoenzyme of phospholipase C) by both phosphorylation and physical recruitment of PLCγ. The phosphorylated tyrosines of the activated receptor form docking sites for the PLCγ SH2 domains, enabling its phosphorylation and relocalization into the proximity of its cell membrane substrate, phosphatidylinositol bisphosphate ($PtdInsP_2$). PLCγ hydrolyzes $PtdInsP_2$ into inositol triphosphate ($InsP_3$) and diacylglycerol (DAG), which are two important intracellular second messengers. InsP3 causes an increase in intracellular calcium and DAG activates many members of the PKC family. Subsequently, a complex set of enzymes mediates the regeneration of multiple inositol phosphates.[1] Several studies indicate that PLCγ is able to trigger a mitogenic signal; for instance, microinjection of purified PLCγ results in DNA synthesis.[37]

Activation of PLCγ can be evaluated by analyzing its increased tyrosine phosphorylation after immunoprecipitation with specfic antibodies. The stimulation of the NIH EGFR/ret cells with EGF induces phosphorylation of PLCγ to an extent comparable to that elicited by activation of EGFR in NIH EGFR cells. The increased phosphorylation of PLCγ results in increased $PtdInsP_2$ breakdown. A two-fold increase of inositol phosphate formation has been observed in EGF-triggered NIH EGFR/ret and EGFR cells compared to untreated cells (Table 3.1).[36] The ret protein contains the sequence YLDL (residues 1015-1018),[4] corresponding to the Y-L/V-E/D-L/I/V consensus involved in the binding of the N-terminal SH2 domain of PLCγ. The presence of this amino acid sequence

Table 3.1. Comparison between the signaling pathways of the RET, PDGFR and EGFR tyrosine kinases in NIH 3T3 cells

Intracellular transducer	RET	EGFR	PDGFR
PLC-γ[1]	0.1	0.1	1
PtdIns-3K[2]	0	0.1	1
ras-GTP[3]	1.6	1.3	1
raf	no	yes	yes
MAPK	no	yes	yes
eps8	no	yes	yes
eps15	no	yes	yes
ezrin	no	yes	yes
paxillin	yes	yes	yes

1 Enzymatic activity measured as inositol phosphate formation and compared to that induced by the PDGFR used as reference.
2 Enzymatic activity measured as generation of radiolabeled phosphatidylinositol 3-phosphate and compared to that induced by the PDGFR used as reference.
3 Ras activation was measured as GTP-/GDP-bound ratio.

in the carboxyterminal tail of ret suggests that tyrosine 1015 might be an autophosphorylation site involved in the binding of PLCγ. Ret or EGFR mediated stimulation of PLCγ in NIH 3T3 cells is not as extensive as the stimulation mediated by the endogenous PDGFR. Treatment of either NIH EGFR or NIH EGFR/ret cells with PDGF-BB resulted in at least a tenfold higher level of PLCγ phosphorylation, and a 5-10 fold increase in inositol phosphate formation (Table 3.1).[41]

Lack of effect of ret activation on PI3-K

One of the first signaling molecules identified as a receptor-associated protein was phosphatidylinositol 3-kinase (PtdIns-3K or PI3-K), an enzyme involved in the metabolism of inositol-containing phospholipids. PI3-K is capable of phosphorylating the D-3 position of the inositol ring of phosphatidylinositol and of its phosphorylated isoforms. PI3-K is a heterodimer consisting of 85-kDa (p85) and 110-kDa (p110) subunits. The catalytic activity resides in the p110 subunit. The p85 subunit contains two SH2 domains, one SH3 domain and a bcr-related domain.[1] The SH2 domains can bind to several autophosphorylated RTKs, allowing p85 to function as an adapter between the PI3-K catalytic subunit and tyrosine kinases.[38,39] There is strong evidence that PI3-K plays a central role in mitogenic signaling. Several mitogenic factors can activate PI-3 kinase, and several oncoproteins (abl, crk, fps and middle T) are associated with PI-3 kinase.[40] Mutants of Middle T antigen or of PDGFR that do not associate with PI3-K do not induce mitogenic effects.[40] PI3-K products activate the γ isoform of PKC, which is important in mitogenic signaling. This suggests that the PI3-K pathway may be mediated by PKCγ.[1]

Ret does not appear to activate the PI3-K pathway. Neither association with nor phosphorylation of the p85 subunit of PI3-K was detected in NIH EGFR/ret cells or in NIH EGFR cells after EGF stimulation.[36] In contrast, p85 phosphorylation was clearly detectable after stimulation of the endogenous PDGF receptor in these cells. Similarly, no PI3-K enzymatic activity was detected in NIH EGFR/ret cells, and only minimal activity was observed in NIH EGFR cells.[36] This result is in accordance with the fact that none of the 18 tyrosines present in the cytoplasmic domain of ret display the consensus sequences Y-M/I/V/E-X-M and Y-M/L/I-X-M, that are thought to bind to the N-terminal and C-terminal SH2 domains, respectively, of PI-3K.[41]

Coupling of RET with the ras mitogenic pathway

The ras gene family includes three members, H-ras, K-ras and N-ras, which all code for proteins of 21 kDa (p21). These proteins possess GTPase activity and they are able to bind to and hydrolyze GTP to GDP and P_i. The GTP-bound form of the Ras protein is active in signal transduction, while hydrolysis of GTP to GDP and P_i leads to the switching off of ras activity. Since very little of the Ras protein in most cells is in the GTP-bound active state, stimuli which can increase the level of ras GTP can activate ras-dependent signaling pathways. Several human and experimental tumors harbor mutated transforming forms of ras. Most of these mutations abrogate the GTPase activity, thus providing the conditions for the continuous presence of Ras in its activated GTP-bound state.[42] Ret, like EGFR, PDGFR, and other receptor tyrosine kinases acts, at least in part, through stimulation of ras. Upon EGF triggering of an EGFR-ret chimera in NIH-3T3 cells, the level of ras GTP was increased, demonstrating that ret, like EGFR and PDGFR, provokes ras activation (Table 3.1).[36] Ret induction of ras function probably has pleiotropic effects. In analogy to other receptors, ret is probably involved in the transduction of mitogenic signals. Oncogenic ret versions (*RET*/PTC1 and *RET*/PTC3) are able to induce meiotic maturation of *Xenopus laevis* oocytes. This effect is mediated by ras signaling, since neutralizing anti-ras antibody blocked ret-induced oocyte maturation without interfering with the accumulation and tyrosine phosphorylation of the ret/PTC proteins.[43] In addition, ras activation is associated with the differentiation effects exerted by ret in PC12 cells (see below).

Tyrosine kinase activation of ras is mediated by guanine nucleotide exchange factors, such as Sos, which stimulate the replacement of GDP by GTP in ras proteins. The Sos exchange factor is activated indirectly by tyrosine kinases. Intracellularly, Sos is complexed with adapter molecules which can recognize and bind specific phosphotyrosine residues in proteins. Upon autophosphorylation of a receptor tyrosine kinase, the adaptor-Sos complex will bind to the tyrosine kinase. By this means, Sos is brought into proximity with membrane-bound ras GDP, thereby allowing stimulation of GDP→GTP exchange and activation of ras. Two adapter molecules have been well characterized. Grb2, which binds directly to Sos, can bind to a specific tyrosine phosphorylated domain (pYXNX) on other proteins, including autophosphorylated tyrosine kinases.[35] Another adapter protein, Shc, binds by interaction of its PTB (phosphotyrosine binding domain) with specific tyrosine phosphorylated sites (NPXpY) distinct from those recognized by SH2 domains. Shc does not directly bind Sos; instead, Shc, activated by tyrosine phosphorylation, binds to Grb2-Sos complexes via the Grb2 SH2 domain.[44] Thus, Sos can be tethered to receptor tyrosine kinases autophosphorylated at either pYXNX or NPXpY sites. It has been demonstrated that oncogenic forms of ret (ret/PTC1 and ret/PTC2) were able to associate with Grb2 as well as tyrosine phosphorylated Shc. Whether the binding of ret to Grb2 is direct or mediated by an interaction with Shc is unknown.[45]

As mentioned above, ret encodes for two protein products differing in their carboxyterminal tail. Nine amino acids in the short isoform are substituted by 51 different amino acids in the "long" isoform.[4] The short isoform is devoid of tyrosine residues corresponding to the consensus binding sites for the Grb2 SH2 domain (pYXNX) or the Shc PTB domain (NPXpY). In the long isoform of ret, two Grb2 SH2 consensus binding sites (corresponding to tyrosine residues at positions 1090 and 1096) are present, suggesting that the two ret isoforms differ in their ability to associate with these adaptors. Nevertheless, Shc and Grb2 form complexes with the short isoform of ret,[45] suggesting either that additional adapters participate in this complex, or that non-consensus phosphotyrosine binding sites on ret are employed.

Activation of ras by ret is also associated with binding of p120GAP to ret, as demonstrated through the use of the EGFR-ret chimera.[36] The association of p120GAP to ret might have direct implications for stimulation of ras function, because p120GAP, when complexed with other receptors, has a reduced ability to promote the conversion of active ras to the GDP-bound inactive form.[46] In addition, p120GAP may act as a shuttle to direct the interaction of ret with p60 and p190 proteins. Although the function of these proteins is still debatable, it is interesting to observe that p190 is a GTPase activating protein (GAP protein) for members of the Rho subfamily of small GTPases.[47] Since Rho proteins are involved in controlling the organization of the actin cytoskeleton, the ret→p120GAP→p190→Rho association could enable ret to transmit signals to the cytoskeleton.

Downstream effectors of ret signal transduction

Control of gene transcription by extracellular signals must involve nuclear translocation of the signal. Some pathways involve migration of transcription factors into the nucleus after their activation in the cytoplasm. In other pathways, proteins that regulate the activity of transcription factors serve as a shuttle for the transmission of the signal from the cytosol to the nucleus, a case in point being mitogen activated protein kinases (MAPKs) that transmit ras-dependent signals (this pathway is reviewed in chapter 4).[48-51]

We investigated activation of raf and MAPK in NIH EGFR/ret by analysing their phosphorylation upon triggering with EGF. Even though ret is able to signal through ras, neither raf nor the MAPK cascade was activated, indicating that ras signals are routed through alternative pathways.[36] Several observations suggest that ras and the raf pathways can be divergent. For instance, activated ras induces the expression in PC 12 cells of certain neurite-specific genes that are not induced by activated raf.[52] In 3T3-L1 cells, oncogenic ras activated MAPK, but oncogenic raf did not.[51] Finally, ras is required for multiple signals induced by v-src in fibroblasts, some of which may be independent of raf.[54] There are several candidates for alternative downstream targets of ras. Certainly, the recently described ras-activatable MEKK/SEK/SAPK and MKK3/p38HOG1 pathways, reviewed in chapter 4, must be considered. In addition, several other potential downstream targets of ras, which may affect other pathways, exist. P120GAP is believed to be a ras effector, because it binds the ras effector region and is necessary for some effects of ras.[55] Another potential candidate is PI3-K, which associates with activated ras.[56] Finally, PKC[57] and phosphatidylcholine-specific phospholipase C[58] are considered to be other potential ras effectors. Interestingly, ret/PTC1 is able to induce malignant transformation of cultured thyroid cells only upon cooperation with activated v-ras (see above).[34] This is a relatively unusual cooperative phenomenon; the cooperation between "nuclear" (such as myc) and "cytoplasmic" (such as ras or a tyrosine kinase) oncogenes being a much more classical case of cooperation in other systems.[59] Because of the poor ability of ret to induce raf and MAPKs, these observations suggest that activation of raf and MAPKs by v-ras could be the biochemical basis for the cooperation between ret/PTC1 and v-ras in the transformation of thyroid cells. Another intriguing observation is the similarity between ret and TSH-R mitogenic pathways. Thyrocytes are a specific target for ret activation by rearrangement in vivo. Growth control in thyrocytes is physiologically controlled by thyrotropin (TSH) through cAMP- and ras-dependent pathways.[60] TSH-R signaling through ras is not associated with the activation of the raf-MAPK cascade,[61] resembling what happens in the case of ret signaling.

The mutated forms of ret associated with MEN2A and MEN 2B have different substrates

As described in chapter 1, MEN 2B is a syndrome similar to MEN 2A, but clinically more aggressive. It is characterized by the absence of parathyroid disease, and the presence of mucosal neuromas, marfanoid habitus and less frequently diffuse intestinal ganglioneuromatosis, thickened corneal nerves, skeletal abnormalities and delayed puberty. The ret mutations of two syndromes differ, but in both cases ret is converted into a dominant oncogene, as shown by transfecting these mutated ret genes into NIH 3T3 cells. In the case of MEN 2A, point mutations causing the substitution of extracellular cysteine residues, result in dimerization and constitutive activation of the ret kinase.[18,19] No dimers were detected in ret-MEN 2B transfectants, suggesting a different mechanism for the activation of ret in MEN 2B.[18] The MEN 2B-associated mutation affects the putative substrate pocket of the catalytic core of ret, by substituting a methionine residue, which is found in most other receptor tyrosine kinases, with a threonine, thus making this region more similar to cytoplasmic tyrosine kinases.[62] This mutation is expected to affect substrate specificity. Indeed, maps of pTyr-containing tryptic peptides of ret-MEN2A and ret-MEN2B proteins showed different autophosphorylation sites, indicating alterations in the autocatalytic specificity caused by the MEN2B mutation.[18] In addition, two-dimensional electrophoretic patterns of phosphorylated intracellular proteins indicate that several proteins are differentially phosphorylated in ret-MEN2A and -MEN2B mutants. This supports the notion that the MEN2B mutation alters substrate specificity of ret.[18]

The specific switch in substrate preference between wild type ret and ret-MEN 2B has recently been elucidated. By using a degenerate peptide library, Cantley's group has shown that cytosolic tyrosine kinases preferentially phosphorylate substrates with Ile or Val at the -1 position and a Glu, Gly or Ala at the +1 position (with respect to the position of phosphorylated tyrosine). In contrast, receptor tyrosine kinases select substrates with Glu at the M1 position and large hydrophobic amino acids at the +1 position. The wild type ret protein preferentially phosphorylates the optimal peptide substrate for the receptor tyrosine kinase EGFR, whereas the MEN2B mutant phosphorylates optimal substrates for src and abl cytoplasmic kinases.[63] Taken together, these results suggest that altered signaling could account for the differences in phenotype seen in the MEN 2B syndrome.

CELLULAR EFFECTS OF RET SIGNALING

Specificity of the ret transduction pathway

Central to the understanding of signal transduction is the question of specificity of response to stimuli, such as ligand-mediated activation of receptor tyrosines kinases. This question is divided into two parts. First, by what mechanism do different cell types respond differently to activation of a specific receptor? Second, how can two receptors, using ostensibly similar signal transduction pathways, induce disparate responses in a single cell? A complete discussion of these problems is beyond the scope of this chapter; numerous reviews of specific systems are available.[1,64-68] Here, the specificity of the cellular response to ret will be reviewed.

The cellular effects of ret signaling have been studied mainly in NIH 3T3 and 32D cells, in which ret signaling is mitogenic, and in PC12 pheochromocytoma cells, in which ret signaling results in differentiation. The results of these studies are described below.

Ret induced mitogenesis in NIH 3T3 and 32D cells

Transfection of EGFR/ret and erb2 suggested that these receptor tyrosine kinases all have similar potency in transformation assays in NIH 3T3 cells.[36] Different results were obtained when the non-tumorigenic myeloid cell line, designated 32D, was used. These cells are devoid of endogenous EGFR and are absolutely dependent on interleukin-3 (IL-3) for their proliferation and survival.[69] Introduction of EGFR into these cells allows them to survive and proliferate in the presence of EGF, even in the absence of IL-3. Unlike EGFR, ret and erbB2 are weak transducers of mitogenic signals in 32D cells. In fact, IL-3 dependence was relieved after EGFR transfection,[70] but not after transfection with ret/PTC1, or the EGFR/ret chimera, or an EGFR-erbB2 chimera.[36] This suggests that in 32D cells, ret and erbB2 may employ different signal transduction effectors than those used by EGFR. Alternatively, the same signal transduction effectors may be employed, but they may be activated to different extents, either quantitatively or temporally. Similar models have been proposed to explain the differences in the response of PC12 cells to NGF, which induces extensive neuronal differentiation, and EGF, which is mitogenic in these cells (see chapter 4).[67]

To investigate the molecular basis for the potent transforming and mitogenic actions of ret, two-dimensional maps of phosphorylated proteins after EGF treatment of NIH EGFR/ret were performed and compared with those obtained with EGFR and ErbB2 expressing cells. Two of the several proteins found phosphorylated, were invariably phosphorylated to higher stoichiometry after ErbB-2 and EGFR/ret activation; these were paxillin and a novel 23 kD protein.[71] Purification of p23 will be required to elucidate its function. Paxillin is a 68-kDa cytoskeletal protein involved in actin-membrane attachment at sites of cell adhesion to the extracellular matrix. Extensive tyrosine phosphorylation of paxillin occurs in src transformed cells during integrin-mediated cell adhesion and during embryonic development.[72] Paxillin physically interacts with vinculin and its close coupling with p125Fak (focal adhesion kinase) suggests that paxillin may be a substrate for p125Fak in vivo. In addition, the ability of paxillin to interact with a variety of intracellular signaling molecules is suggested by domains within its sequence. Indeed, analysis of the paxillin sequence has revealed the presence of five tyrosine-containing sequences that conform to SH2-binding motifs for src, crk, PLC-γ and p85 PI3-K, and of a proline-rich region indicative of an SH3-binding domain. In fact, paxillin was shown to bind to the SH2 domains of v-crk and Csk and to the SH3 domain of c-src.[73] All these observations suggest that paxillin could be a protein involved in the regulation of tyrosine kinase activity by integrin engagement and that tyrosine phosphorylation of paxillin can interfere with actin-cytoskeleton assembly.

Differences between EGFR and ret signaling pathways were also observed in the weak ability of ret to phosphorylate the recently identified EGFR substrates eps8 and eps15. Eps8 is an evolutionary conserved protein of 97-kDa containing an SH3 domain. It appears to be involved in mitogenic control because it becomes phosphorylated, after activation of the EGFR and several other tyrosine kinase receptors and its overexpression enhances mitogenic signaling.[74,75] The eps15 gene product migrates as a 142-155-kDa doublet. Eps15 activity may be relevant to the processes regulating cell proliferation because this protein is able to transform NIH 3T3 cells.[76,77] We found that both eps8 and eps15 were very weakly phosphorylated after ret activation. Activation of the ErbB2 kinase was totally ineffective in inducing eps15 phosphorylation but did result in eps8 phosphorylation.[36] Also ezrin, a protein that belongs to a family of molecules with homology to the 4.1 protein band, involved in establishing connection between the cytoskeleton and the plasma membrane, is efficiently phosphorylated by EGFR and PDGFR[78] but not upon ret activation.[36] Together, these data suggest differences in downstream effectors of ret and EGFR signaling.

RET INDUCES A DIFFERENTIATION PATHWAY IN PC12 CELLS

The association of ret expression with tissues of neural crest derivation and the involvement of ret point mutations in different neurocristopathies (such as MEN 2 syndromes and Hirschsprung's disease) (see above) prompted a study of ret signaling in neural-crest derived cells. As described in chapter 4, PC12 cells are the most widely used in vitro model system for neural crest differentiation. These rat pheochromocytoma cells show a chromaffin-like phenotype that shifts to resemble sympathetic neurons upon NGF stimulation.[79] Conversely, a mitogenic but not a differentiation effect is exerted on PC12 cells by insulin and EGF.[78] Activated ret isoforms (ret/PTC1 and ret/PTC3) were able to induce a differentiation response in PC12 cells. Transcription driven by NGF-inducible gene promoters, such as NGFI-A and vgf, belonging to primary or delayed NGF response genes, and from neuron specific enolase is rapidly induced in PC12 cells by these activated forms of ret.[81] These observations support the notion that ret signaling, in vivo, is involved in the establishment of the differentiated phenotype of neural-crest derived cell lineages. In this system, ret appears to utilize at least part of the ras dependent signal transduction pathway induced by NGF in PC12 cells (see chapter 4). NGF had no additional effect on the activity of ret-induced NGFI-A or vgf promoters in PC12 cells. In addition, ras activity is required to enable ret to induce this differentiative response because the dominant negative mutant ras p21N17 blocks these effects.[81]

FUTURE PERSPECTIVES

The primary function of receptor tyrosine kinases is to convert extracellular information (represented by growth factors) into chemical signals that can be relayed into the cell. Upon interaction with the specific ligand, the tyrosine-phosphorylated receptor or one of its substrates associates with intracellular signaling enzymes acting as effectors. Receptor tyrosine kinases are involved in signaling both cell proliferation and differentiation. Their role in mitogenesis is indicated by the frequent isolation of oncogenic variants from human or experimental tumors.[82] A role in determining cell differentiation is well-exemplified by studies of developmental expression of RTKs and their targeted disruption, and by genetic studies in *Drosophila melanogaster* and *Caenorhabditis elegans*.[81] Ret plays a central role in growth and differentiation control in specific tissues. Oncogenic isoforms of ret are generated by gene rearrangement events in papillary thyroid carcinomas and germ-line or somatic point mutations in neural crest derived tumors. Recent data suggest that the highly transforming activity of these ret oncogenes is mediated by pathways not completely overlapping with those of other well-studied receptors such as EGFR and PDGFR. The isolation of novel ret substrates is still in progress. However, a central role in the mitogenic effects of ret seems to be played by p21ras, although its signal does not seem to be routed through the classical raf-MAPK pathway. Activation of endogenous ras is also important for the differentiation effects of ret in PC12 cells. Together with studies on the developmental expression of ret and on the effects of ret gene knock-out, the described effects of ret in PC12 cells support the idea that a ligand for ret is very likely a molecule involved in the embryonal development of the neural and excretory systems.

Acknowledgments

This work was supported by grants from the AIRC and the CNR, "Progetto Finalizzato ACRO." The authors are indebted to Dr. V. de Franciscis and to all the members of the research group directed by Prof. G. Vecchio for their contribution.

References

1. Kazlauskas A. Receptor tyrosine kinases and their targets. Current Opinion in Genetics and Development 1994; 4:5-14.
2. Bishop JM. Molecular themes in oncogenesis. Cell 1991; 64:235-48.
3. Takahashi M, Ritz J, Cooper GM. Activation of a novel human transforming gene, ret, by DNA rearrangement. Cell 1985; 42:581-8.
4. Takahashi M, Buma T, Iwamoto Y et al. Cloning and expression of the ret proto-oncogene encoding a tyrosine kinase with two potential transmembrane domains. Oncogene 1988; 3:571-8.
5. Iwamoto T, Taniguchi M, Asai N et al. cDNA cloning of mouse ret proto-oncogene and its sequence similarity to the cadherin superfamily. Oncogene 1993; 8:107-91.
6. Schuchardt A, Srinivas S, Pachnis V et al. Isolation and characterization of a chicken homolog of the c-ret proto-oncogene. Oncogene 1995; 10:641-9.
7. Sugaya R, Ishimaru S, Hosoya T et al. A *Drosophila* homolog of human proto-oncogene ret transiently expressed in embryonic neu-

ronal precursor cells including neuroblasts and CNS cells. Mechanisms of Development 1994; 45:139-45.
8. Pachnis V, Mankoo B, Costantini F. Expression of the c-RET protooncogene during mouse embryogenesis. Development 1993; 119:1005-17.
9. Avantaggiato V, Dathan NA, Grieco M. et al. Developmental expression of the RET protooncogene. Cell Growth and Diff 1994; 5:305-11.
10. Tsuzuki T, Takahashi M, Asai N et al. Spatial and temporal expression of the ret proto-oncogene product in embryonic, infant and adult rat tissues. Oncogene 1995; 10:191-8.
11. Fabien N, Paulin C, Santoro M et al. Expression of the RET protooncogene in normal human C-cells and adrenal medulla. Int J Onc 1994; 4:623-6.
12. Fabien N, Paulin C, Santoro M et al. The RET proto-oncogene is expressed in predominantly epithelial human thymomas. Int J Onc 1994; 5:489-93.
13. Schuchardt A, D'Agati V, Larsson-Blomberg L et al. Defects in the kidney and enteric nervous system of mice lacking the tyrosine kinase receptor ret. Nature 1994; 367:380-3.
14. Romeo G, Ronchetto P, Luo Y et al. Point mutations affecting the tyrosine kinase domain of the RET proto-oncogene in Hirschsprung's disease. Nature 1994; 367:377-8.
15. Edery P, Lyonnet S, Mulligan LM et al. Mutations of the RET proto-oncogene in Hirschsprung's disease. Nature 1994; 367:378-80.
16. Santoro M, Rosati R, Grieco M et al. The ret proto-oncogene is consistently expressed in human pheochromocytomas and thyroid medullary carcinomas. Oncogene 1990; 5:1595-8.
17. Ikeda I, Ishizaka Y, Tahira T et al. Specific expression of the ret proto-oncogene in human neuroblastoma cell lines. Oncogene 1990; 5:1291-6.
18. Santoro M, Carlomagno F, Romano A et al. Activation of RET as a dominant transforming gene by germline mutations of MEN2A and MEN2B. Science 1995; 267:381-3.
19. Asai N, Iwashita T, Matsuyama M et al. Mechanism of activation of the *ret* proto-oncogene by multiple endocrine neoplasia 2A mutations. Mol Cell Biol 1995; 15:1613-9.
20. Fusco A, Grieco M, Santoro M et al. A new oncogene in human thyroid papillary carcinomas and their lymph-nodal metastases. Nature 1987; 328:170-2.
21. Bongarzone I, Pierotti MA, Monzini N et al. High frequency of activation of tyrosine kinase oncogenes in human papillary thyroid carcinoma. Oncogene 1989; 4:1457-62.
22. Grieco M, Santoro M, Berlingieri MT et al. PTC is a novel rearranged form of the ret proto-oncogene and is frequently detected in vivo in human thyroid papillary carcinomas. Cell 1990;

60:557-63.
23. Santoro M, Carlomagno F, Hay ID et al. RET oncogene activation in human thyroid neoplasms is restricted to the papillary carcinoma subtype. J Clin Invest 1992; 89:1517-22.
24. Grieco M, Cerrato A, Santoro M et al. Cloning and characterization of H4 (D10S170), a gene involved in RET rearrangements in vivo. Oncogene 1994; 9:2531-5.
25. Pierotti MA, Santoro M, Jenkins RB et al. Characterization of a chromosome 10q inversion juxtaposing RET and H4 genes and creating the oncogenic sequence PTC. Proc Natl Acad Sci USA 1992; 89:1616-20.
26. Bongarzone I, Monzini N, Borrello MG et al. Molecular characterization of a thyroid tumor-specific transforming sequence formed by the fusion of ret tyrosine-kinase and the regulatory subunit RI alpha of cyclic AMP protein kinase A. Mol Cell Biol 1993; 13:358-66.
27. Santoro M, Dathan NA, Berlingieri MT et al. Molecular characterization of RET/PTC3: a novel rearranged version of the RET proto-oncogene in a human thyroid papillary carcinoma. Oncogene 1994; 9:509-16.
28. Bongarzone I, Butti MG, Coronelli S et al. Frequent activation of the ret proto-oncogene by fusion with a new activating gene in papillary thyroid carcinomas. Cancer Res 1994; 54:2979-85.
29. Lanzi C, Borrello MG, Bongarzone I et al. Identification of the product of two oncogenic rearranged forms of the RET proto-oncogene in papillary thyroid carcinomas. Oncogene 1992; 7:2189-94.
30. Santoro M, Sabino N, Ishizaka Y et al. Involvement of RET oncogene in human tumors: specificity of RET activation to thyroid tumors. Br J Cancer 1993; 68:460-4.
31. Iwamoto T, Takahashi M, Ito M et al. Oncogenicity of the ret transforming gene in MMTV/ret transgenic mice. Oncogene 1990; 5:533-42.
32. Iwamoto T, Takahashi M, Ito M et al. Aberrant melanogenesis and melanocytic tumor development in transgenic mice that carry a metallothionein/ret fusion gene. EMBO J 1991; 10:3167-75.
33. Fusco A, Berlingieri MT, Di Fiore PP et al. One and two-step transformation of rat thyroid epithelial cells by retroviral oncogenes. Mol Cell Biol 1987; 7:3365-70.
34. Santoro M, Melillo RM, Berlingieri MT et al. The TRK and RET tyrosine-kinase oncogenes cooperate with ras in the neoplastic transformation of a rat thyroid epithelial cell line. Cell Growth and Diff 1993; 4:77-84.
35. Schlessinger J. SH2/SH3 signaling proteins. Current Biology 1994; 4:25-30.
36. Santoro M, Wong WT, Aroca P et al. An epidermal growth factor

receptor/*ret* chimera generates mitogenic and transforming signals: evidence for a *ret*-specific signaling pathway. Mol Cell Biol 1994; 14:663-75.
37. Majerus PW, Ross TS, Cunningham TW et al. Recent insights in phosphatidylinositol signaling. Cell 1990; 63:459-65.
38. Escobedo JA, Kaplan DR, Kavanaugh WM et al. A phosphatidylinositol-3 kinase binds to platelet-derived growth factor receptors through a specific receptor sequence containing phosphotyrosine. Mol Cell Biol 1991; 11:1125-32.
39. Hu P, Margolis B. Skolnik R et al. Interaction of phosphatidylinositol-3 kinase-associated p85 with epidermal growth factor and platelet-derived growth factor receptors. Mol Cell Biol 1992; 12:981-90.
40. Whitman M, Kaplan DR, Schaffhausen B et al. Association of phosphatidylinositol kinase activity with polyoma middle T competent for transformation. Nature 1985; 315:239-42.
41. Songyang Z, Shoelson SE, Chauduri G et al. SH2 domains recognize specific phosphopeptide sequences. Cell 1993; 72:767-78.
42. McCormick F. Activators and effectors of *ras* p21 proteins. Current Opinion in Genetics and Development 1994; 4:71-6.
43. Grieco D, Dathan NA, Santoro M et al Activated RET oncogene products induce maturation of *Xenopus* oocytes. Oncogene 1995; 11:113-7.
44. Pelicci G, Lanfrancone L, Grignani F et al. A novel transforming protein (SHC) with an SH2 domain is implicated in mitogenic signal transduction. Cell 1992; 70:93-104.
45. Borrello MG, Pelicci G, Arighi E et al. The oncogenic versions of the RET and TRK tyrosine kinases bind shc and Grb2 adaptor proteins. Oncogene 1994; 9:1661-8.
46. Serth J, Weber W, French M et al. Binding of the H-ras p21 GTPase activating protein by the activated epidermal growth factor receptor leads to inhibition of the p21 GTPase activity *in vitro*. Biochemistry 1992; 31:6361-5.
47. Nobes C, Hall A. Regulation and function of the Rho subfamily of small GTPases. Current opinion in genetics and development 1994; 4:77-81.
48. Herskowitz I. MAP kinase pathways in yeast: for mating and more. Cell 1995; 80:187-97.
49. Lange-Carter CA, Johnson GL. Ras-dependent growth factor regulation of MEK kinase in PC12 cells. Science 1994; 265:1458-61.
50. Roberts TM. A signal chain of events. Nature 1992; 360:534-5.
51. Marshall CJ. MAP kinase kinase kinase, MAP kinase kinase and MAP kinase. Current Biology 1994; 4:82-9.
52. D'Arcangelo G, Halegoua S. A branched signaling pathway for nerve growth factor is revealed by Src-, Ras-, and raf-mediated gene inductions. Mol Cell Biol 1993; 13:3146-55.
53. Porras A, Muszynski, Rapp UR et al. Dissociation between activa-

tion of Raf-1 kinase and the 42-kDa mitogen-activated protein kinase/90-kDA S6 kinase (MAPK/RSK) cascade in the insulin/Ras pathway of adipocytic differentiation of 3T3 L1 cells. J Biol Chem 1994; 269:12741-8.
54. Qureshi SA, Alexandropoulos K, Rim M et al. Evidence that Ha-Ras mediated two distinguishable intracellular signals activated by v-Src. J Biol Chem 1992; 267:17635-9.
55. Polakis P, McCormick F. Interactions between p21ras proteins and their GTPase activating proteins. Cancer Surv 1992; 12:25-42.
56. Rodriguez-Viciana P, Warne PH, Dhand R et al. Phosphatidylinositol-3-OH kinase as a direct target of Ras. Nature 1994; 370:527-32.
57. Berra E, Diaz-Meco MT, Dominguez I et al. Protein kinase isoform is critical for mitogenic signal transduction. Cell 1993; 74:555-63.
58. Cai H, Erhardt P, Troppmair J et al. Hydrolysis of phosphatidylcholine couples Ras to activation of Raf protein kinase during mitogenic signal transduction. Mol Cell Biol 1993; 13:7645-51.
59. Hunter T. Cooperation between oncogenes. Cell 1991; 64:249-70.
60. Vassart G, Dumont JE. The thyrotropin receptor and the regulation of thyrocyte function and growth. Endocr Rev 1992; 13:596-611.
61. Al-Alawi N, Rose DW, Buckmaster C et al. Thyrotropin-induced mitogenesis is ras dependent but appears to bypass the raf dependent cytoplasmic kinase cascade. Mol Cell Biol 1995; 15:1162-8.
62. Hofstra RMW, Landsvater RM, Ceccherini I et al. A mutation in the RET proto-oncogene associated with multiple endocrine neoplasia type 2B and sporadic medullary thyroid carcinoma. Nature 1994; 367:375-76.
63. Songyang Z, Carraway III KL, Eck MJ et al. Catalytic specificity of protein-tyrosine kinases is critical for selective signaling. Nature 1995; 373:536-40.
64. van der Geet P, Hunter T, Lindberg RA. Receptor protein-tyrosine kinases and their signal transduction pathways. Ann Rev Cell Biol 1994; 10:251-337.
65. Ben-Baruch N, Yarden Y. Neu differentiation factors: a family of alternatively spliced neuronal and mesenchymal factors. Proc Soc Exp Biol Med 1994; 206:221-7.
66. Gherardi E, Sharpe M, Lane K et al. Hepatocyte growth factor/scatter factor (HGF/SF), the c-met receptor and the behavior of epithelial cells. Symp Soc Exp Biol 1993; 47:163-81.
67. Marshall CJ. Specificity of receptor tyrosine kinase signaling: transient versus sustained extracellular signal-regulated kinase activation. Cell 1995; 80:179-85.
68. Hill CS, Treisman R. Transcriptional regulation by extracellular signals: mechanisms and specificity. Cell 1995; 80:199-211.
69. Greenberger JS, Sakakeeny MA, Humphries RK et al. Demonstration of permanent factor-dependent multipotential (erythroid/neu-

trophil/basophil) hematopoietic progenitor cell lines. Proc Natl Acad Sci USA 1983; 80:2931-5.
70. Pierce JH, Ruggiero M, Fleming TP et al. Signal transduction through the EGF receptor transfected in IL-3 dependent hematopoietic cells. Science 1988; 239:628-31.
71. Romano A, Wong WT, Santoro M et al. The high transforming potency of erbB-2 and ret is associated with phosphorylation of paxillin and a 23 kDa protein. Oncogene 1994; 9:2923-33.
72. Turner CE. *Paxillin* is a major phosphotyrosine-containing protein during embryonic development. J Cell Biol 1991; 115:201-7.
73. Turner CE, Miller JT. Primary sequence of *paxillin* contains putative SH2 and SH3 domain binding motifs and multiple LIM domains: identification of a vinculin and pp125Fak-binding region. J Cell Sci 1994; 107:1583-91.
74. Fazioli F, Minichiello L, Matoska P et al. Eps8, a substrate for the epidermal growth factor receptor kinase, enhances EGF-dependent mitogenic signals. EMBO J 1993; 12:3799-808.
75. Wong WT, Carlomagno F, Druck T et al. Evolutionary conservation of the EPS8 gene and its mapping to human chromosome 12q23-q24. Oncogene 1994; 9:3057-61.
76. Fazioli F, Minichiello L, Matoskova B et al. *eps15*, a novel tyrosine kinase substrate exhibits transforming activity. Mol Cell Biol 1993; 13:5814-28.
77. Wong WT, Kraus MH, Carlomagno F et al. The human eps 15 gene, encoding a tyrosine kinase substrate, is conserved in evolution and maps to 1p31-p32. Oncogene 1994; 9:1591-7.
78. Bretscher A. Rapid phosphorylation and reorganization of ezrin and spectrin accompany morphological changes induces in A-431 cells by epidermal growth factor. J Cell Biol 1989; 108:921-30.
79. Greene LA, Tischler AS. Establishment of a noradrenergic clonal line of rat adrenal phaeochromocytoma cells that respond to nerve growth factor. Proc Natl Acad Sci USA 1976; 76:2424-8.
80. Ohmichi M, Pang L, Ribon V et al. Divergence of signaling pathways for insulin in PC12 pheochromocytoma cells. Endocrinology 1993; 133:46-56.
81. Califano D, Monaco C, De Vita G et al. Activated RET/PTC oncogene elicits immediate early and delayed response genes in PC12 cells. Oncogene 1995; 11:107-12.
82. Pawson T, Hunter T. Signal transduction and growth control in normal and cancer cells. Current Opinion in Genetics and Development 1994; 4:1-4.

CHAPTER 4

Mechanisms of Growth Factor-Mediated Signal Transduction in PC12 Cells

Michael P. Myers, Kenneth D. Swanson and Gary Landreth

One of the most important and pressing issues in developmental neurobiology is understanding how neurons are generated from uncommitted precursor populations and how they acquire the specific phenotypic characters which distinguish them as mature neurons. As described in chapter 5, there has been substantial progress in understanding the complex intercellular relationships and the role of neurotrophins and other growth factors during nervous system development. These studies have relied principally on the use of in vivo models and primary tissue culture systems. While illuminating, these studies are of limited utility in the investigation of the intracellular, biochemical mechanisms subserving the actions of growth factors that direct the differentiation of these precursor cells into neurons. These populations of precursor cells are reliant upon growth factors for their survival, and this dependency confounds the design of experiments to explore the mechanism of the growth factors actions. Moreover, the limited amount of material available for study and the absence of appropriate control populations of cells has impeded the progress in elucidating

Genetic Mechanisms in Multiple Endocrine Neoplasia Type 2, edited by Barry D. Nelkin.

the cellular mechanisms responsible for the effects of growth factors. An especially interesting problem is found in the sympathoadrenal lineage of the neural crest, in which the partially committed sympathoadrenal precursor cells retain the plasticity to differentiate into either mature adrenal chromaffin cells or sympathetic neurons. This lineage is the precursor, in MEN 2, for pheochromocytoma development. These considerations led Greene and Tischler to approach this problem by developing clonal cell lines derived from tumors which could be maintained in tissue culture yet retain responsiveness to factors which would provoke their differentiation into a neuronal phenotype. These studies resulted in the isolation of the clonal PC12 cell line derived from a rat pheochromocytoma[1] that would proliferate in an undifferentiated state and resembled in many ways the embryonic neural crest cells, which are the progenitors of both mature adrenal chromaffin cells and sympathetic neurons.[2] The PC12 cells have the remarkable ability to differentiate into a phenotype that closely resembles that of mature sympathetic neurons in response to NGF.[1] Upon treatment with NGF, PC12 cells exit the cell cycle and begin to develop a variety of neuronal attributes, including the elaboration of neurites, the development of electrical excitability and the expression of a variety of neuron-specific genes.[3] Similar to sympathetic neurons, differentiated PC12 cells become dependent on NGF for their survival, as removal of NGF results in the subsequent death of a substantial fraction of these cells.[4] The development of this model system has permitted detailed studies of the intracellular signal transduction mechanisms and the biochemical processes required for neuronal differentiation. The PC12 cells have become one of the most extensively used cell culture models for signal transduction and have been essential to the progress in understanding nervous system development at a molecular level. As described in chapter 3, this cell culture model is now being used as a model for alterations in signal transduction in MEN 2.

Although the response of PC12 cells to NGF is the best studied, a number of other factors have been identified that induce PC12 cells to acquire a neuronal phenotype, including fibroblast growth factor (FGF) and interleukin-6.[3] Paradoxically, many of the signals generated by differentiating stimuli, such as NGF or FGF, are remarkably similar to those that are generated in response to mitogens, such as epidermal growth factor (EGF) or insulin.[5]

Surprisingly, a number of genes which have been implicated in controlling cellular proliferation are also believed to be necessary for PC12 cell differentiation, and the expression of a number of oncogenes results in the morphological differentiation of these cells.[6]

The objective of this review is to describe, in outline, the biochemical events that are initiated upon growth factor binding to PC12 cells. These events mediate the transmission of signals from the membrane to cytoplasmic effectors and ultimately to transcription factors in the nucleus. These signaling pathways are highly complex and interactive reflecting the capacity of growth factors to modulate the most fundamental aspects of cellular functions, such as metabolic rate, survival and differentiation. It is important to point out that our knowledge of these events is fairly recent and the rapid pace at which the details of signal transduction pathways are being revealed makes it certain that our present understanding of these events is rudimentary.

NGF RECEPTORS

The response of PC12 cells to NGF is initiated upon binding of NGF to its cell surface receptors. Historically, early work from the laboratories of Shooter and Bradshaw revealed that NGF bound to its target cells with two different affinities, 90% of which were characterized as low affinity receptors ($K_d = 10^{-9}$ m) and approximately 10% of which were high affinity receptors ($K_d = 10^{-11}$ m).[7-9] Occupancy of high affinity receptors is correlated with the induction of biological responses. In 1986, the low affinity NGF receptor was cloned[10,11] and is referred to as p75, indicative of its molecular weight. p75 is a 75 kDa transmembrane protein with an extracellular ligand binding domain and a short cytoplasmic domain having no known catalytic activity (Fig. 4.1A).[11] p75 appears to be promiscuous with respect to its interactions with the other neurotrophins, binding all species with equal affinity.[12] The function of this molecule remains ambiguous. There is growing evidence, however, that p75 acts to transduce intracellular signals and has been implicated in the initiation of apoptosis.[13] Recently, NGF binding to p75 has been shown to activate sphingomyelinase, resulting in the generation of the lipid second messenger ceramide.[14] This effect is similar to the action of the TNFα receptor, a structural homologue of p75, although the biological actions of the ceramide second messenger are not well-described.

The identification of the trkA proto-oncogene as a receptor for NGF provided the key to deducing how NGF initiates signal transduction cascades within its target cells.[15-20] The trkA gene product is comprised of three functional domains: a large extracellular ligand binding domain, a single transmembrane domain and a cytoplasmic domain, which encodes a tyrosine specific protein kinase (Fig. 4.1B). NGF binding to trkA activates the intrinsic tyrosine kinase activity of the receptor, which is ultimately responsible for generating intracellular signals. TrkA is structurally and functionally similar to other growth factor receptors in this regard, and exerts its actions through its ability to phosphorylate proteins on tyrosine residues.[21] In contrast to p75, trkA has been shown to be indispensable for NGF signaling.[22,23] TrkA binds specifically to NGF, while other closely related receptor tyrosine kinases, i.e., trkB and trkC, are receptors for the other neurotrophins (BDNF and neurotrophin 4/5 for trkB, neurotrophin 3 for trkC).[24,25] The structural conservation of the kinase domains of the trk receptors (around 90%) suggests that their intracellular actions are likely to be similar; however, differences in the biological effects of these receptors have been observed (see chapter 8).

Nature of the High Affinity NGF Receptor

One might have thought that the identification of two NGF receptors would have led to the rapid resolution of the nature of the high affinity NGF receptor. However, this has not proven to be the case, and a vigorous controversy surrounds this problem. Two basic models have been advanced; in the first, p75 and trkA interact to form heterodimers, or, alternatively, trk monomers interact to form homodimers which are proposed to constitute the biologically active high affinity receptor (Fig. 4.1C).[15,16,19]

In the first model, the high affinity NGF receptor has been hypothesized to be comprised of trkA-p75 hetrodimers. Support for this model comes from studies in which the expression of either p75 or trkA in COS cells results in the formation of only low affinity binding sites.[15] When p75 is coexpressed with trkA,

Fig. 4.1. (on page 65) NGF receptor models. NGF has been shown to bind two cell surface receptors. The low affinity p75 receptor (A) and the trkA proto-oncogene (B). It has been proposed that the high affinity binding site consists of either trkA homodimers or trkA-p75 heterodimers (C).

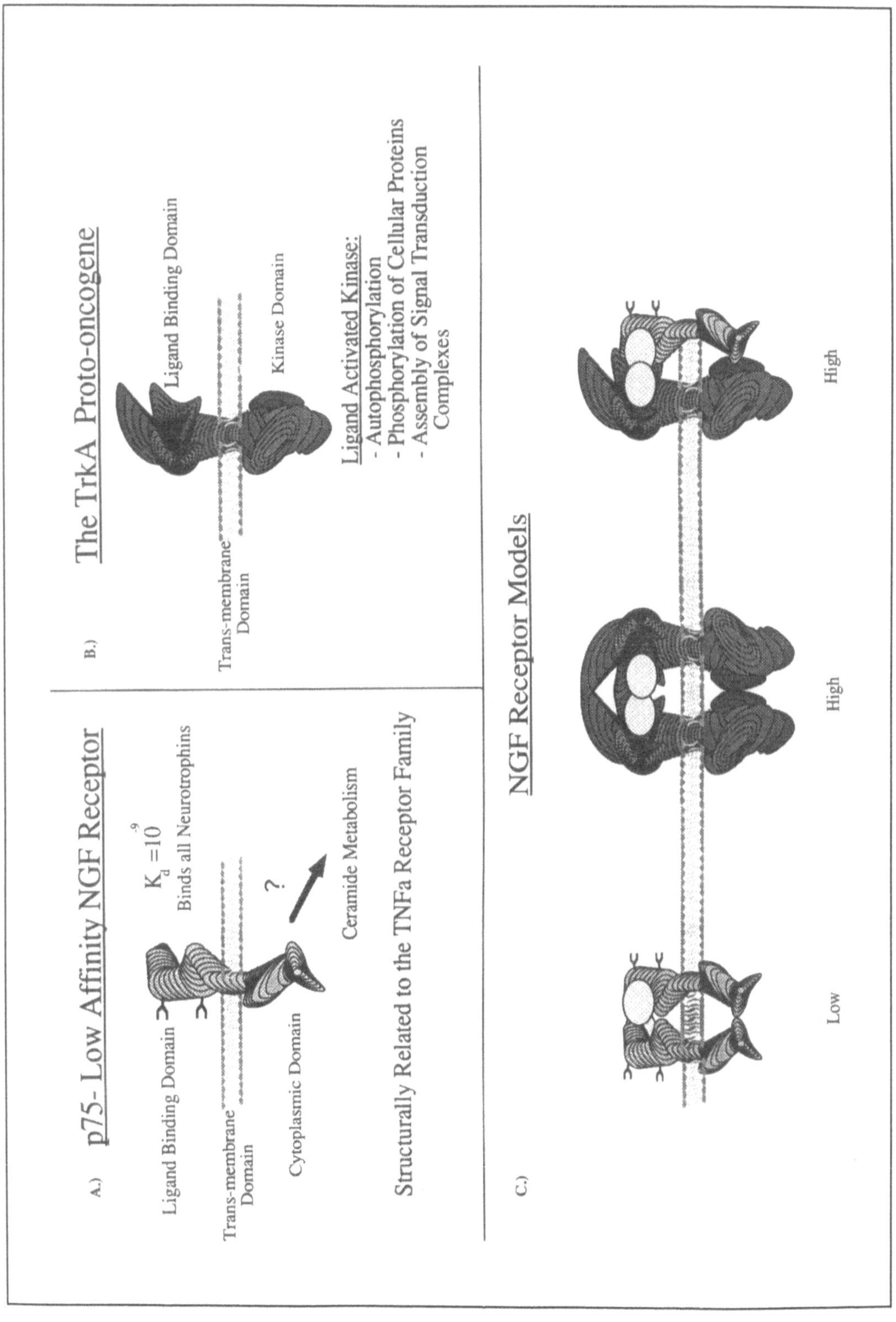
A.)
p75- Low Affinity NGF Receptor
Ligand Binding Domain
Trans-membrane Domain
Cytoplasmic Domain
$K_d = 10^{-9}$
Binds all Neurotrophins
?
Ceramide Metabolism
Structurally Related to the TNFa Receptor Family
B.)
The TrkA Proto-oncogene
Ligand Binding Domain
Trans-membrane Domain
Kinase Domain
Ligand Activated Kinase:
- Autophosphorylation
- Phosphorylation of Cellular Proteins
- Assembly of Signal Transduction Complexes
C.)
NGF Receptor Models
Low
High
High

however, high affinity binding sites are formed.[15] Similar results were found in NR18 cells, a variant PC12 cell line that expresses a mutant, non-functional form of p75.[26] These cells no longer bind NGF with high affinity, resulting in cells that are unresponsive to NGF.[26] Expression of wild-type p75 in NR18 cells, however, restores high affinity NGF binding and NGF responsiveness.[15,17,26] The observation that p75 and trkA bind to different regions of NGF has been suggested to play a role in the formation of receptor heterodimers and makes this model conceptually appealing, although there is no direct evidence that this is the case. More recently, it has been proposed that p75 acts to "present" NGF to trkA and may account for the exquisite sensitivity of its target neurons to low levels of NGF.[27] Treatment of PC12 cells with either a mutant form of NGF that does not interact with p75 or with an anti-p75 antibody that blocks NGF binding to p75, reduces NGFs ability to stimulate intracellular signaling events.[27] Moreover, p75 appears to potentiate cellular responses to all of the neurotrophins.[28]

It has also been proposed that the biologically relevant form of the NGF receptor is comprised of trkA homodimers. In this model, trkA acts in a manner analogous to other growth factor receptor kinases. Ligand binding provokes the activation of the intrinsic tyrosine kinase activity of the receptor and receptor dimerization, resulting in intermolecular autophosphorylation of trkA. Receptor autophosphorylation on tyrosine residues then triggers the formation of a large signaling complex centered around the activated receptors. Support for this model derives from studies in which trkA was expressed in 3T3 cells or Xenopus oocytes (cells that do not normally express detectable levels of p75), resulting in the formation of high affinity binding sites for NGF.[19,20] Expression of p75 in the absence of trkA resulted in the formation of only low affinity binding sites.[19,20] Importantly, treatment of trkA-expressing 3T3 cells with NGF resulted in the stimulation of mitosis and ultimately the transformation of these cells.[19] Expression in PC12 cells of chimeric receptors comprised of the TNFα ligand binding domain and trkA transmembrane and cytoplasmic domains resulted in the differentiation of the cells in response to TNFα.[29] Similar chimeras of the TNFα receptor binding domain and the intracellular domain of p75 receptors, on the other hand, showed no response to TNFα.[29] Perhaps the best evidence for the trkA homodimer model comes from mice in which the p75 receptor has been genetically inactivated.[22] Mice deficient in p75 exhibit a number of neuronal abnormalities, especially those involving sensory neurons. Surprisingly, p75 knock-out mice have essentially normal sympathetic ganglia, a tissue which is highly sensitive to perturbations in NGF signaling.[22] These data suggest that, although p75 is important for proper development of sensory neurons, it is not essential for NGF signaling in sympathetic neurons.

The conflicting nature of the data makes it impossible to rule out either model. Perhaps these issues will be clarified once a specific function for p75 is found. In either case, stimulation of trkAs tyrosine kinase activity is absolutely necessary for NGF signaling.

INTRACELLULAR SIGNALING MECHANISMS

Extracellular stimuli, such as growth factors, act through the dissemination of signals from the plasma membrane to intracellular targets. These signals are carried by a battery of growth factor-stimulated signaling pathways. These signaling events are transmitted by traditional second messenger systems as well as more recently described pathways involving the recruitment of a constellation of proteins into complexes whose assembly is initiated by binding of the ligand to the growth factor receptor. The specific components recruited into these complexes are ultimately responsible for shaping the cellular response. These receptor-associated complexes then initiate a variety of downstream events that eventually result in a cellular response. Although a number of signaling pathways are activated in response to NGF, signaling through the GTP-binding protein p21ras has been shown to be an essential event mediating the morphological differentiation of PC12 cells. Signals leading to the activation of p21ras and its downstream effectors will therefore be discussed in detail.

FORMATION OF SIGNALING COMPLEXES WITH GROWTH FACTOR RECEPTORS

Perhaps the single most important event following growth factor treatment is the autophosphorylation of the growth factor receptors. The newly formed phosphotyrosine residues become docking sites for proteins possessing SH2 domains.[30] SH2 domains were first recognized as a conserved motif found in a number of signaling molecules, and so named src-homology domains (SH2), due to the initial discovery of this motif within the src family of tyrosine kinases.[30] SH2 domains are absolutely critical for growth factor signaling as they bind directly to phosphotyrosine residues.[31] The specificity of the interactions between specific SH2 domains and phosphotyrosine residues is determined by the amino acids surrounding the phosphorylated tyrosine residues and mediates the association of these signaling molecules with the receptor.[31] The diversity of the cellular response to growth factor stimulation is governed by the SH2 domain-bearing proteins that interact with the tyrosine phosphorylated receptor.

Two major classes of SH2 domain proteins have been identified. One class of SH2 domain proteins has an associated catalytic activity such as the ras GTPase activating protein, p120GAP, phospholipase Cγ (PLCγ), phosphoinositol 3-kinase (PI3-K), the src family of tyrosine kinases and even some tyrosine phosphatases.[30] A second class of proteins consists of small SH2 domain-bearing proteins that act as molecular adapters and mediate the formation of signaling complexes associated with the activated receptors.[30] These proteins often contain multiple SH2 domains and a related motif termed an SH3 domain.[30] SH3 domains mediate protein-protein interactions by binding to proline-rich motifs present in other proteins.[32] These SH2-SH3 domain adapter proteins are able to bind to multiple proteins function as molecular adapters by binding to the tyrosine phosphorylated receptor via their SH2 domains and to other effector molecules via their other src homology domains.[32]

A subset of SH2 domain-containing proteins have been shown to associate directly with trkA.[33-36] TrkA has been shown to be phosphorylated principally on three tyrosine residues, tyr490, tyr751 and tyr785.[33-36] The tyr490 residue is the binding site for the adaptor protein shc, tyr785 is the binding site for PLCγ and tyr751 binds PI3-kinase (Fig. 4.2).[33,36] Both PLCγ and PI3-kinase rapidly become tyrosine phosphorylated and activated in response to NGF treatment.[33,36] The activation of PLCγ produces the second messengers inositol triphosphate and diacylglycerol. Inositol triphosphate mobilizes intracellular Ca^{2+} stores, which together with diacylglycerol activate the classical protein kinase C isoforms. PI3-kinase generates phosphatidylinositol 3-phosphates. The specific function of this second messenger is unclear, but its formation is necessary for the activation of the serine/threonine kinase pp70S6K.[37] In a number of cell types, activation of pp70S6K is

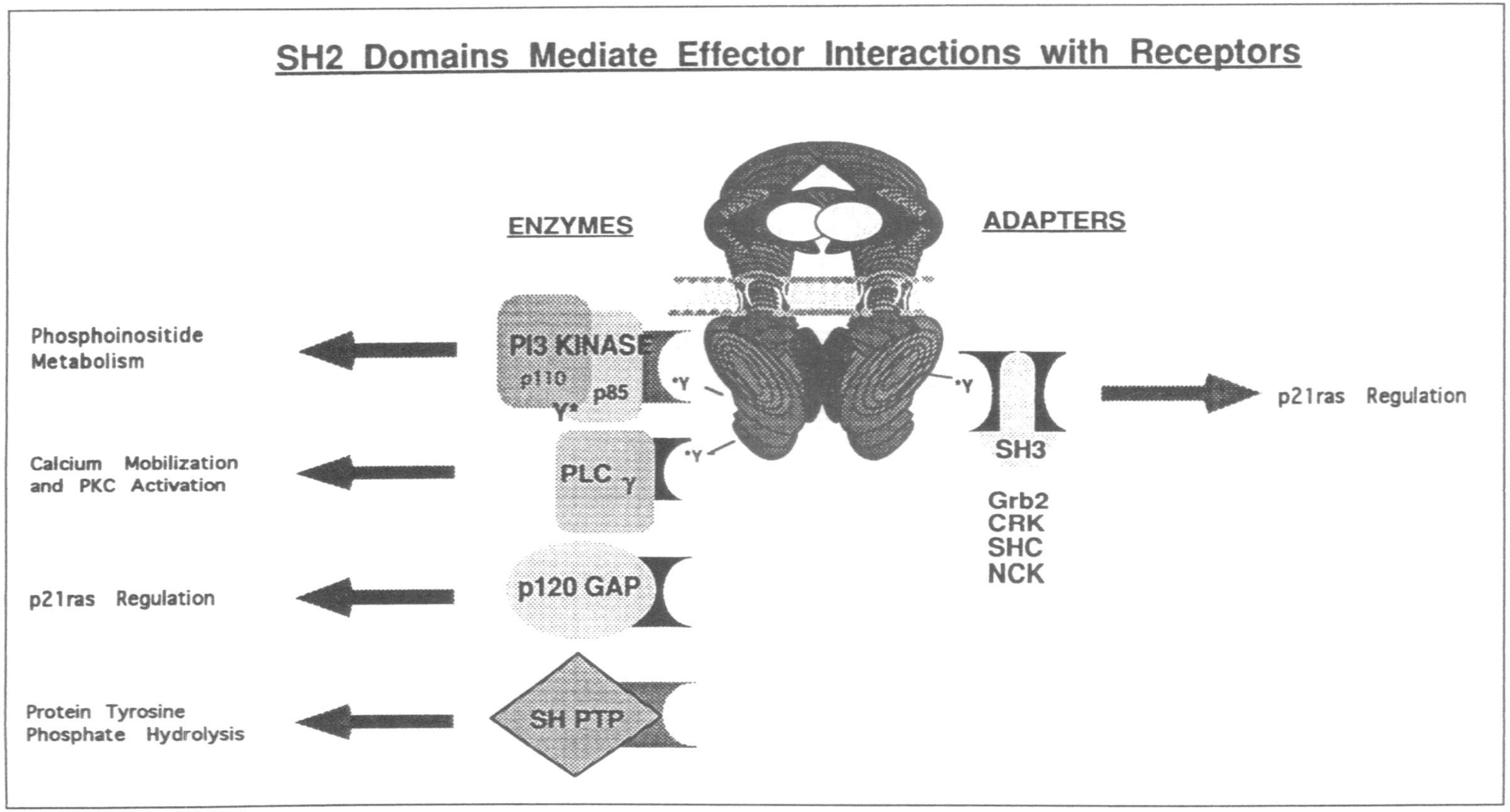

Fig. 4.2. SH2 domains mediate effector interaction with trkA. There are two general classes of SH2 domains, those with an associated enzymatic activity (shown on the right) and those that act as molecular adapters (shown on the left). To date, three SH2 domain-bearing proteins have been shown to interact with trkA, PI3-kinase, PLCγ and Shc.

necessary for cell cycle progression and regulation of translation,[38,39] but the specific function of this enzyme in PC12 cells is unknown.

Shc is an SH2 and SH3 domain-bearing adapter protein that is tyrosine phosphorylated in response to NGF.[40,41] The tyrosine phosphorylation of shc results in the creation of a binding site for another SH2-domain protein, Grb2 (Fig. 4.3).[42-44] The importance of Grb2 rests in its role in the activation of the G-protein, p21ras. Grb2 exists in a pre-formed complex with the guanine nucleotide exchange factor, son of sevenless (Sos).[42,45,46] The ultimate effect of the tyrosine phosphorylation of Shc is the formation of a complex resulting in the translocation of Sos from the cytoplasm to the plasma membrane, where it interacts with p21ras (Fig. 4.3). Sos catalyzes the release of GDP and the binding of GTP to p21ras, converting p21ras into an active form (Fig. 4.4).[47,48] p21ras is inactivated by GTPase activating proteins, such as p120GAP and neurofibromin-1 (Fig. 4.4). These proteins act to stimulate the intrinsic GTPase activity of p21ras.[49,50] A number of recent studies have demonstrated that signaling through Shc is the major avenue leading to the EGF or NGF-mediated activation of p21ras. Peptides that block the interactions between trkA and Shc, or Shc and Grb2, prevent the activation of p21ras in response to EGF and NGF.[40] Similarly, point mutations in trkA that block Shc binding to the receptor also block NGF-stimulated differentiation of PC12 cells.[35,51] Single tyrosine mutations in trkA that disrupt PLCγ or PI3 kinase binding fail to disrupt growth factor-stimulated differentiation of PC12 cells in response to NGF.[51] An interesting distinction in the use of these types of signaling molecules is that Grb2 can directly associate with the EGF receptor, but is unable to directly bind to trkA and requires Shc as an intermediate in the formation of this complex. Surprisingly, stimulation of mutant trkA receptors that bind only PLCγ results in the modest differentiation of PC12 cells.[51] This modest response may indicate the convergence of PLCγ-dependent and Shc-dependent signaling elements. The MAP kinases are a likely site for such convergence, as their activation is p21ras dependent and in some cases PKC dependent.[52,53]

The tyrosine kinase, src, has been suggested to play an important role in NGF signaling, although it has not been shown to bind directly to trkA. Expression of v-src results in the morphological differentiation of PC12 cells, and injection of anti-src

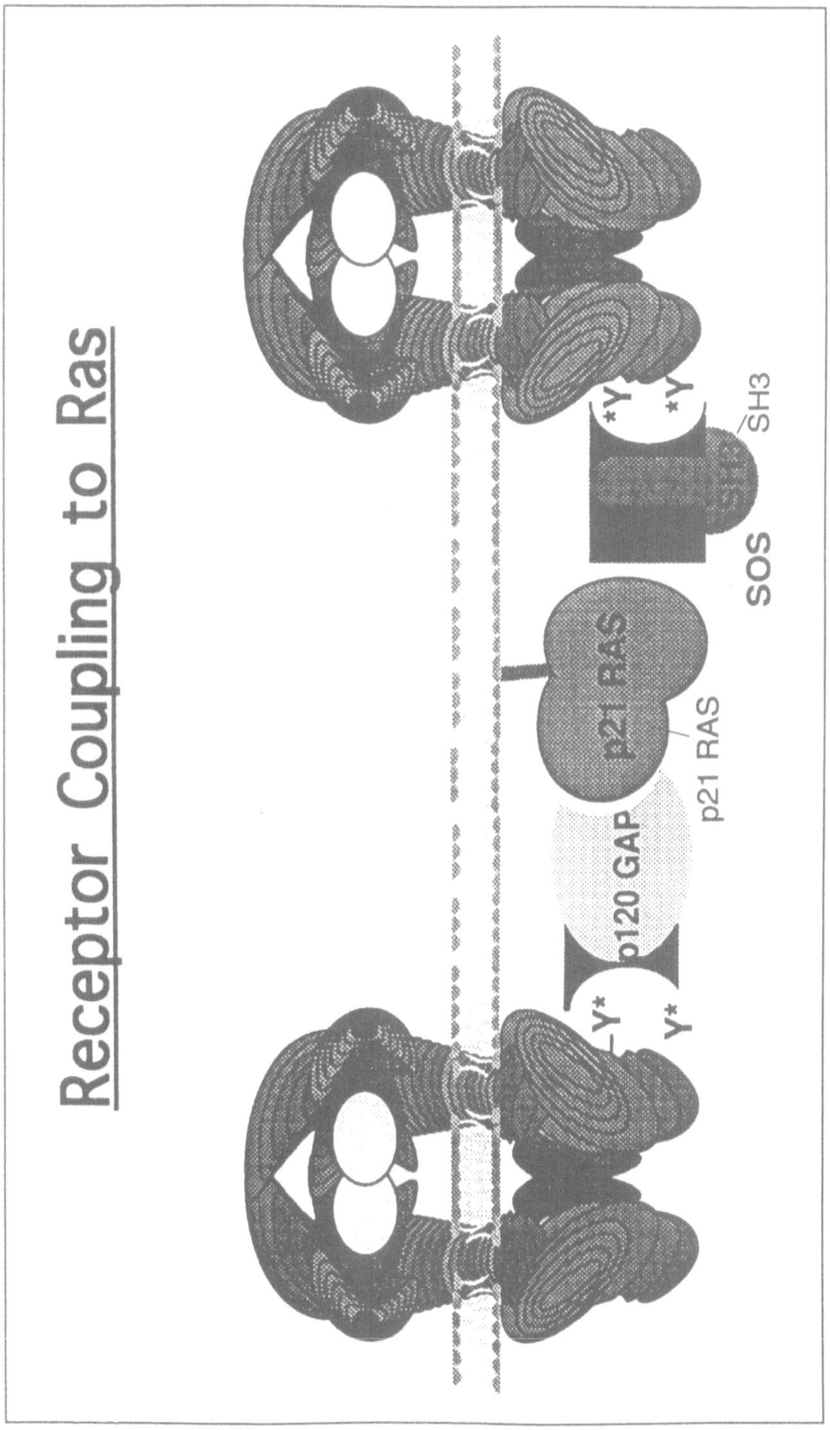

Fig. 4.3. TrkA receptor coupling to p21ras. The binding of NGF to TrkA results in autophosphorylation, creating binding sites for SH2 domain bearing proteins that regulate p21ras and mediate the translocation of these proteins from the cytoplasm to the plasma membrane.

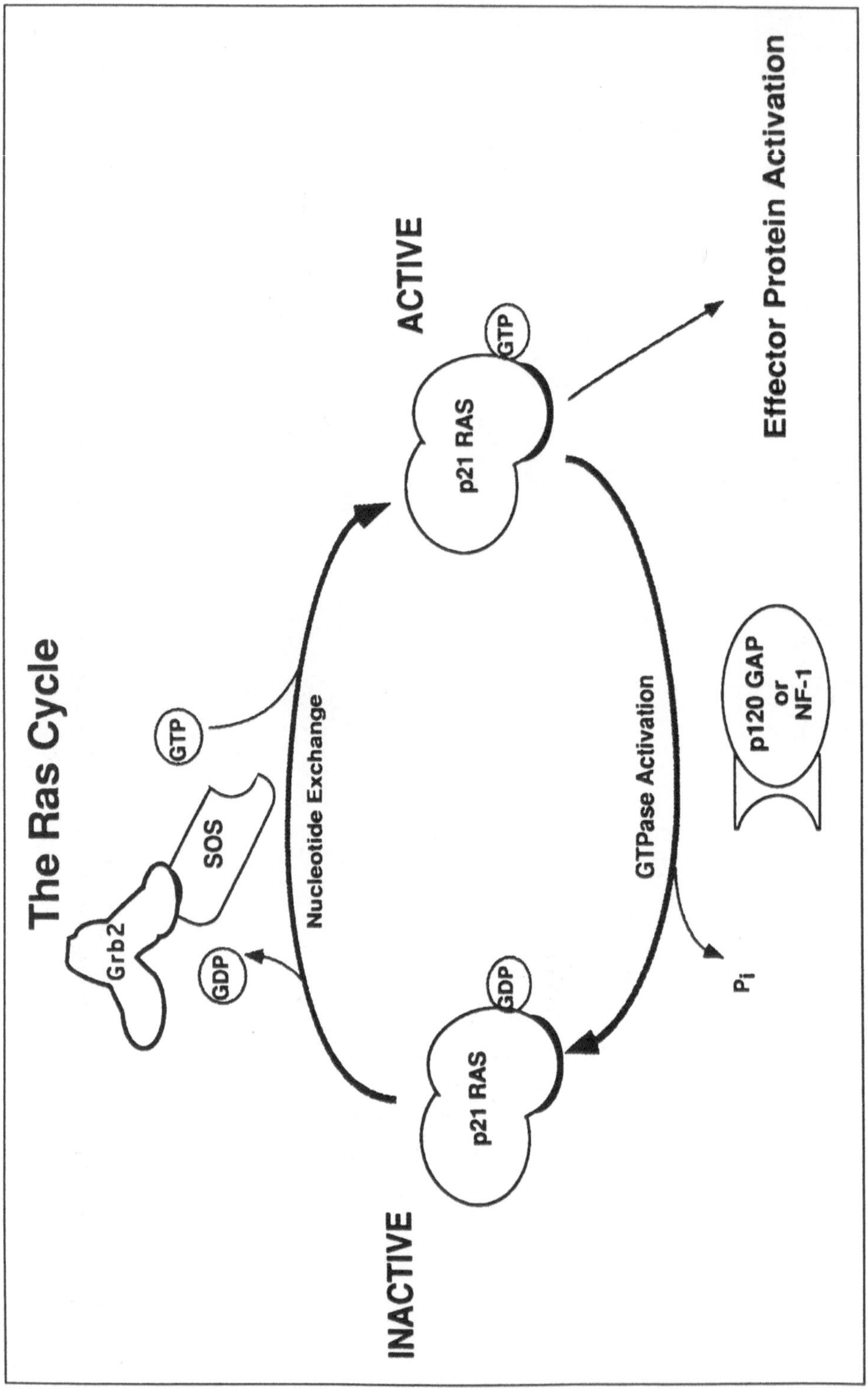

Fig. 4.4. Regulation of p21ras. The conversion of p21ras from an inactive to active state is controlled by guanine nucleotide exchange factors, such as SOS, as well as GTPase activating proteins, such as p120CAP and NF-1. These proteins become physically associated with p21ras at the cell membrane.

antibodies blocks NGF-mediated neurite outgrowth.[54,55] A number of studies have been performed to determine the position of this kinase in the NGF signal transduction pathways. The anti-src antibodies were shown to have no effect on neurite outgrowth elicited in response to activated forms of p21ras.[55] Expression of a dominant negative allele of p21ras (p21ras-N17),[56] on the other hand, blocks v-src mediated neurite outgrowth.[55] These studies indicate that src plays a role in NGF action and is positioned between trkA and p21ras in this pathway. These data suggest that src is likely to be important for the stimulation of p21ras-dependent signaling events.[55]

Regulatory Role of p21ras

In most cell types the activation of p21ras is believed to be necessary for mitogenic signaling. In PC12 cells, however, p21ras signaling is required for differentiation.[57] Expression of oncogenic forms of p21ras in PC12 cells results in the elaboration of extensive neurites.[58] More significantly, expression of a dominant negative allele of p21ras blocks PC12 cell differentiation in response to NGF.[57] Dominant negative mutations of p21ras block the action of NGF and other growth factors by arresting the transmission of signals to downstream components of the cascade.[56] Specifically, expression of dominant negative forms of p21ras blocks NGF-stimulated activation of a number of cytoplasmic kinases, especially the members of the MAP kinase cascade.[59,60] This blockade interferes with the signal transmission between the plasma membrane and the nucleus.[59,60] Importantly, NGF-induced expression of neuron-specific gene products, such as sodium channels and Thy-1, still occurs even when a dominant negative form of p21ras is expressed, indicating that p21ras is not essential for all the effects of NGF and that independent and parallel signal transduction pathways exist to carry information between the plasma membrane and the nucleus.[61]

The MAP Kinase Cascade

The growth factor-stimulated signal transduction pathways discussed so far have been distinguished by two mechanistically distinct processes initiating signal transduction cascades: phosphotyrosine-based assembly of signaling complexes on the receptor and signaling through the GTP-binding protein p21ras. The mechanisms responsible for activating downstream components are different and involve the serial activation of protein kinases which comprise a linear cascade whose targets are both cytosolic and nuclear proteins (Fig. 4.5). The regulated phosphorylation of proteins is the central mechanism through which protein function is regulated and information relayed. The utilization of kinase cascades to transduce biological signals allows for amplification of the initial signal generated upon growth factor binding to their target cells. The MAP kinase cascade allows for the amplification of this signal to levels sufficient to alter gene expression and cellular phenotype. The MAP kinase cascade, as it is currently understood, is shown in Figure 4.5. The activation of the individual elements of this pathway is achieved through the direct phosphorylation of each enzyme by its immediate upstream activator. The activation of the enzymes is transient, as befitting efficient signaling systems, and they are inactivated through the action of protein phosphatases.

The MAP kinase cascade plays a critical role in both mitogenic and differentiative signaling.[52,62,63] In PC12 cells, the MAP kinase cascade is initiated by the exchange of GDP for GTP in p21ras.[40] Binding of GTP opens up the effector domain of p21ras so that it can interact with other proteins, specifically the Raf family of kinases.[64-71] The N-terminal domains of the Raf kinases can then associate with the effector domain of p21ras.[64,72] The importance of the interaction between p21ras and Raf family members is the translocation of Raf to the plasma membrane where it is activated by an undefined mechanism.[73,74] Association with p21ras appears to be necessary, but not sufficient to activate Raf.[69,71] The requirement for p21ras can be bypassed by targeting Raf directly to the plasma membrane,[73,74] or by creating N-terminal deletions of Raf which results in a constitutively active kinase.[75]

Once activated, Raf family members then phosphorylate and activate the dual specificity kinase MEK.[69,76-79] MEK kinase, sometimes referred to as the MAPK kinase, is phosphorylated on serine residues by Raf which results in the activation of MEK. Similar to Grb2 and Sos, it appears that MEK and Raf may exist in preformed complexes.[80] Upon activation, MEK is released from Raf into the cytoplasm[65] where it then phosphorylates the MAP kinases on threonine and tyrosine residues, resulting in the enzymatic activation of these enzymes.[81] The last member of the cascade is pp90rsk, which is phosphorylated and activated by the MAP

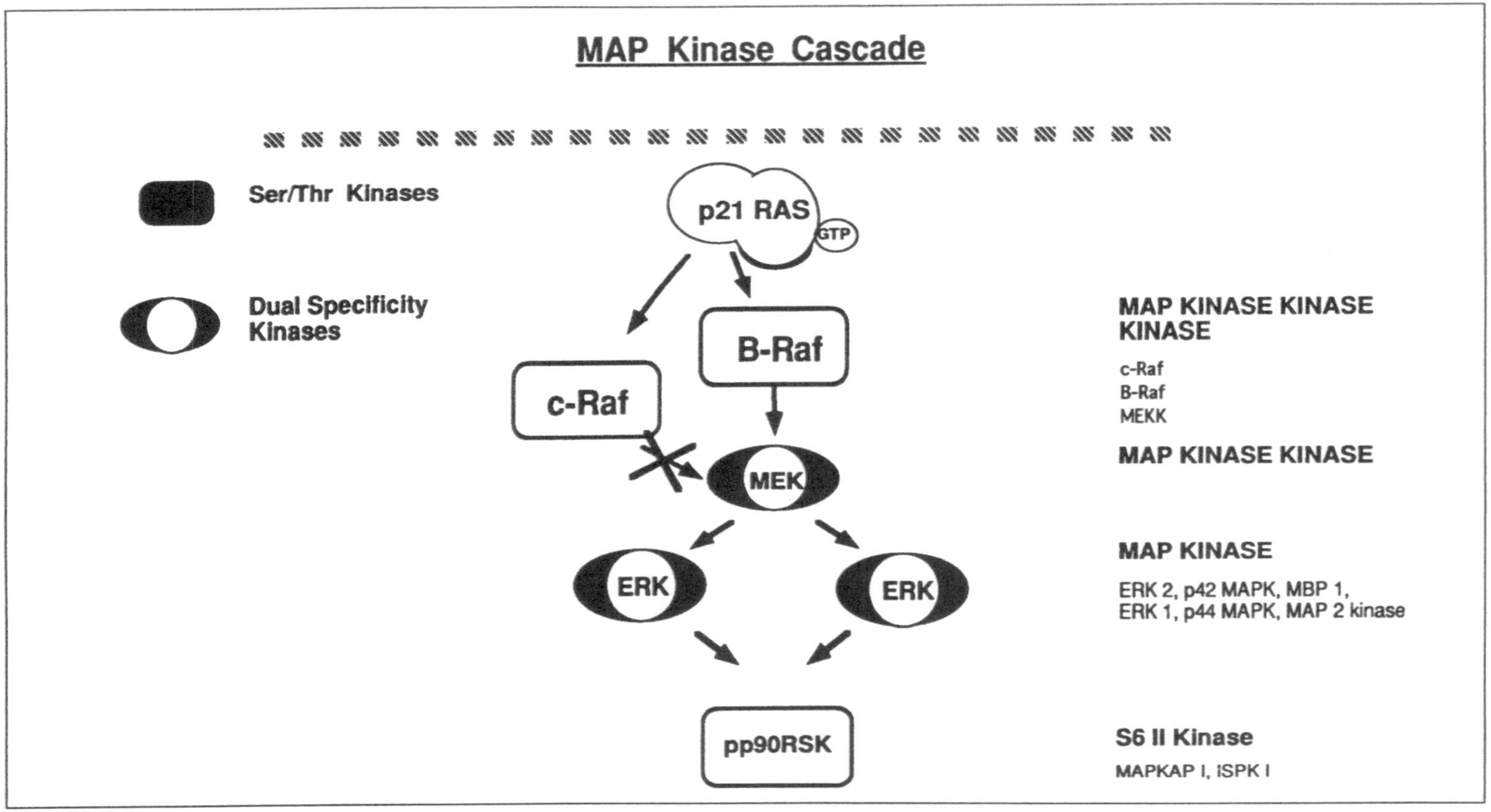

Fig. 4.5. The MAP kinase cascade. In PC12 cells, the direct downstream effector of p21ras is B-Raf, which physically interacts with p21ras. Once activated, B-Raf phosphorylates and activates MEK which in turn activates the MAP kinases. pp90RSK, the last member of the MAP kinase cascade, is directly activated by the MAP kinases. Upon activation, both the MAP kinases and pp90RSK accumulate in the nucleus where they have been shown to phsophorylate transcription factors necessary for immediate early gene expression. C-raf does not activate the MAP kinase cascade in response to NGF in PC12 cells.

kinases.[82-84] Shortly after NGF treatment, the MAP kinases and pp90rsk begin to accumulate in the nucleus. Interestingly, both enzymatically active and inactive forms of the MAP kinases undergo translocation, suggesting that the intracellular transport mechanisms do not rely upon the activity of the enzymes.[85-87]

PC12 cells express two isoforms of Raf, B-Raf and c-Raf.[88,89] Treatment of PC12 cells with NGF results in the activation of only B-Raf.[69] The activity of c-Raf is not stimulated; however, this molecule becomes phosphorylated and exhibits an altered electrophoretic mobility upon NGF treatment.[88,89] It is unclear why c-Raf is not utilized in PC12 cells. It has recently been shown that the association between c-Raf and p21ras can be blocked by the N-terminal phosphorylation of c-Raf by PKA.[90,91] This negative control by phosphorylation of c-Raf, however, is not responsible for blocking c-Raf activation, as both c-Raf and B-Raf isolated from PC12 cells are able to bind to p21ras.[69] B-Raf isolated from PC12 cells, in contrast to c-Raf, is present within a large multiprotein complex of approximately 300-kDa whose constituents are not well-characterized.[69] Interestingly, in other cell types that utilize c-Raf to drive the MAP kinase cascade, c-Raf is also found in large multi-protein complexes.[92] This high molecular weight complex may function to facilitate the translocation of these enzymes from the cytoplasm to the plasma membrane.

The finding that Raf activity is regulated at multiple levels indicates the importance of the MAP kinase pathway for growth factor signaling. As mentioned above, a number of studies has suggested that the stimulation of PKA by forskolin or cAMP analogs may function in PC12 cells to negatively regulate the Raf kinases.[79,93] This finding is surprising, given that cAMP analogs act synergistically, rather than antagonistically, with NGF in the formation of neurites and result in the activation of the MAP kinases.[94] The function of PKA and cAMP in NGF signal transduction is controversial, and the elucidation of the molecular mechanisms subserving NGF action should clear up this controversy.

In almost all cases, stimuli that differentiate PC12 cells also activate the MAP kinases[95]. Recent evidence suggests that the MAP kinases are both necessary and sufficient to differentiate PC12 cells.[95] Expression of a dominant negative MEK allele blocks neurite outgrowth, while expression of activated MEK in PC12 cells results in the formation of neurites.[96] Importantly, neurite formation in response to MEK expression is blocked by co-expression of a kinase-inactive form of ERK2, indicating that MEK acts via the MAP kinases.[96] Similar studies using NIH3T3 cells revealed that MEK is also necessary for cellular transformation, indicating that stimulation of this pathway can result in dramatically different cellular phenotypes and provides a clear demonstration that cell context is an essential determinant of the nature of the cellular response and the phenotype expressed in response to growth factors.[96]

Other MAP Kinase-like Signaling Pathways

It has recently become clear that the organization of the MAP kinase cascade is common to other signal transduction pathways. These newly described pathways are mechanistically similar, and in some cases, they act in parallel with the MAP kinase cascade to transduce signals in response to different classes of stimuli.[97] Recently, a novel MEK activator has been described and termed MEK kinase (MEKK).[98] MEKK is able to phosphorylate and activate MEK in vitro and in vivo when overexpressed in cells.[98] Despite its name, MEKK is apparently not a physiological activator of MEK, and is thought to act principally in one of the parallel MAP kinase-like pathways (Fig. 4.6). The physiological target of MEKK is a MEK homolog, termed SEK.[99] SEK, like MEK, is a dual specificity kinase.[100] SEK phosphorylates and activates the stress activated protein kinases (SAPKs), which are homologues of the MAP kinases. The study of this new family of enzymes has led to a confusing nomenclature, since the SAPKs have been shown to phosphorylate the N-terminus of c-jun and have been termed JNKs (jun N-terminal kinases).[100] The upstream activators of the SAPKs, including SEK, are stimulated by stressful stimuli such as UV radiation, heat shock, protein synthesis inhibitors and inflammatory cytokines, but not generally by growth factor stimulation.[97] This parallel MAP kinase-like cascade, or module, is also p21ras-dependent.[101] The SAPK pathway has been shown to be efficiently activated upon treatment of PC12 cells with the cytokine TNFα, but is stimulated only modestly by NGF and EGF.[97]

Recently, yet another MAP kinase-like module has been discovered that is expressed in PC12 cells and is stimulated by sodium arsenite or osmotic stress (Fig. 4.6).[102] In this case, p38HOG1

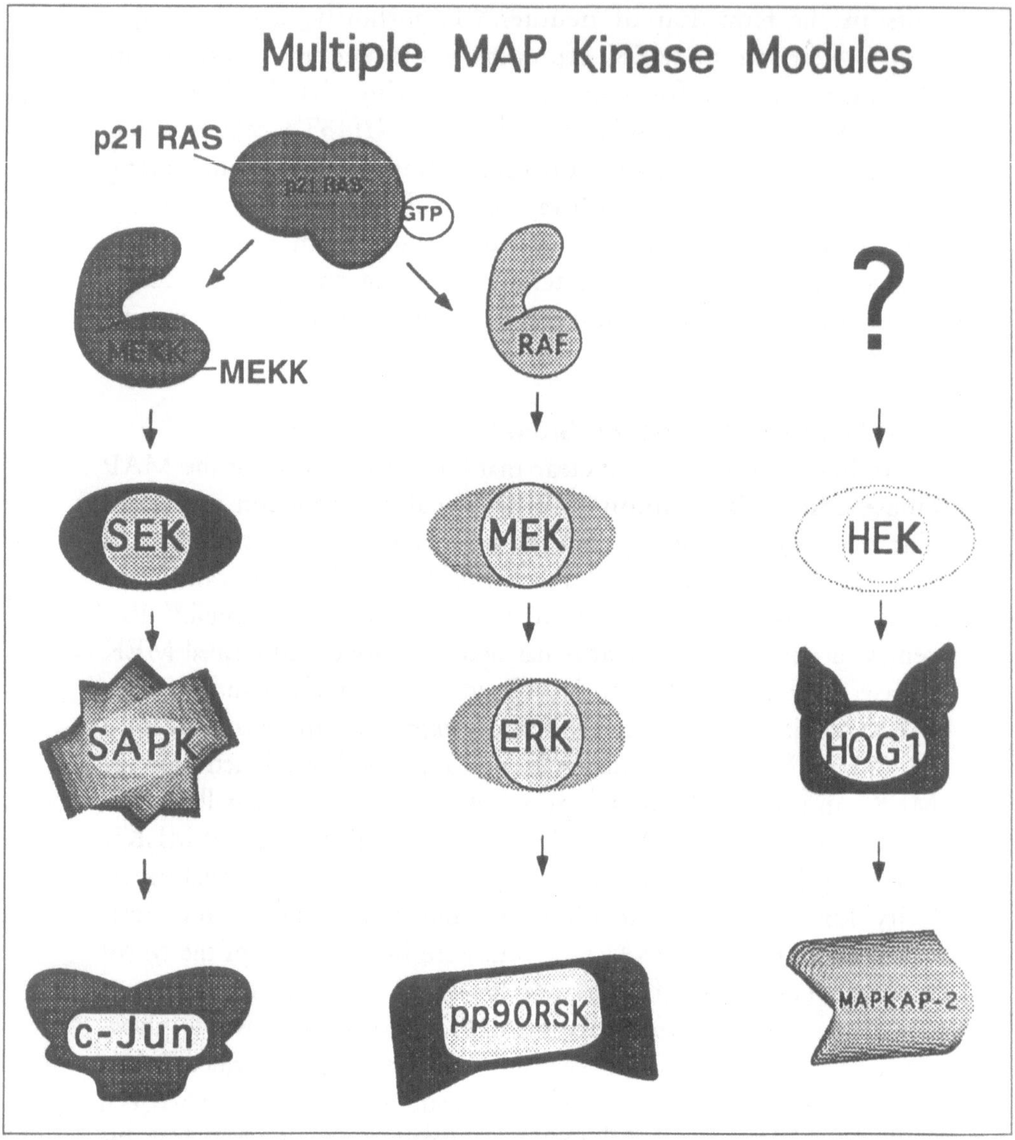

Fig. 4.6. Multiple MAP kinase modules. In addition to the archetypal MAP kinase pathway, several closely related pathways regulating responses to stress and various other stimuli have recently been identified. The best described of these, the stress activated protein kinase (SAP kinase) pathway, is activated by various stressors including protein synthesis inhibitors, ultraviolet radiation and free oxygen radicals. HOG1, yet another MAP kinase homologue, has been shown to be activated by osmotic shock, sodium arsenate and Il-2. HOG1 is activated by HEK, a MEK/SEK family member, and once activated HOG1 serves to activate MAPKAP2.

is the MAP kinase homolog and MKK3 (also called HEK) is the MEK homolog.[103] The upstream activators of this pathway are currently unknown. Downstream of HOG1 is a protein kinase, MAPKAP kinase-2.[102-104] A number of additional MEK, MAP kinase and pp90rsk homologues have been isolated using a variety of techniques. The function of these other homologues is not known, but they are likely to function in other, as yet uncharacterized, MAP kinase-like modules.

It is presently unclear how PC12 cells distinguish between signaling through the multiple MAP kinase-like modules, especially since MEKK and B-Raf are both p21ras-dependent and both are responsive to growth factors. Although p21ras activation is a necessary component, p21ras is clearly not itself capable of activating the Raf kinases. This will probably hold true for MEKK as well. The mechanism(s) determining which signaling module becomes activated is most likely determined by the action of ancillary factors, such as 14-3-3, the membrane associated Raf activator, and PKA.[73,93,105]

Not surprisingly, members of the MAP kinase cascade play a critical role in the regulation of gene expression that occurs in response to NGF.[52,63] In unstimulated cells, the MAP kinases and one of their downstream targets, pp90rsk, are found in both the cytoplasm and the nucleus.[52,85,86] After stimulation, the MAP kinases and pp90rsk are translocated to the nucleus.[52,85,86] Nuclear localization, however, is not dependent on MAP kinase activity as kinase-inactive MAP kinase mutants localize to the nucleus following growth factor stimulation.[86] Once in the nucleus, the MAP kinases and pp90rsk phosphorylate a number of transcription factors, altering their function.[52,63] It is unclear whether the MAP kinases are actively translocated into the nucleus or simply become retained in the nucleus by interacting with their nuclear substrates.

REGULATION OF GENE EXPRESSION

The morphological differentiation of PC12 cells requires the transcription and translation of new gene products.[2] Once these products have accumulated, neurite outgrowth can occur even in the presence of transcriptional and translational inhibitors. These data indicate that morphological differentiation follows two phases, an early transcription-dependent phase and a late transcription-independent phase.[2] The changes in gene expression seen in response to NGF also follows two phases, an early phase occurring in the first hour following NGF exposure and a late phase occurring several hours to days following NGF exposure (Table 4.1). Genes induced during the early phase are largely transcription factors and other regulatory proteins, while genes induced in the late phase are principally structural proteins whose expression is necessary to maintain a neuronal phenotype, as well as those required for the acquisition of electrical excitability.[2]

The induction of early genes in response to NGF occurs in the presence of translational inhibitors, indicating that the induction of these genes occurs as a result of post-translational modifications (such as regulated phosphorylation) of pre-existing proteins.[106] Because these genes are induced in the presence of translation inhibitors, they have been termed immediate early genes by analogy to viral genes exhibiting similar regulation.[106] The transcription factors induced during the early phase of gene expression are believed to act in the control of the expression of the late genes whose expression is blocked by translational inhibitors, and are linked to the acquisition of a neuronal phenotype.[107-110]

Table 4.1. Examples of genes undated by NGF in PC12 cells

Immediate Early genes	**Function**
c-fos	transcription factor
c-jun	transcription factor
MKP-1 (CL100, 3CH134)	Map kinase phosphatase
c-myc	transcription factor
NGFIA (egr-1)	transcription factor
NGFIB (nur77)	transcription factor
Late Genes	
Na+-Channels (type PN1 and type II)	electrical excitability
N-Cam	cell adhesion molecule
Peripherin	intermediate filament protein
SCG10	vesicle-associated protein
Thy-1	extracellular glycoprotein
Transin	protease
Tyrosine Hydroxylase	neurotransmitter synthesis
VGF	neuropeptide

C-FOS AS A MODEL OF GROWTH FACTOR-MEDIATED GENE EXPRESSION

NGF regulates the expression of a variety of genes; however, the best studied of these is the proto-oncogene c-fos. The induction of this immediate early gene will be used here as a model of early gene regulation, which is representative of other growth factor-regulated genes. c-fos is a proto-oncogene which directly binds DNA and acts as a transcriptional regulator. c-fos mRNA is virtually undetectable in unstimulated PC12 cells; however, upon growth factor treatment of PC12 cells, there is a rapid induction of c-fos mRNA. c-fos mRNA reaches maximal levels within 30 minutes and returns to unstimulated levels within 60 minutes following NGF treatment.[106,111,112] Transcription of c-fos is induced by a variety of extracellular stimuli, including NGF, EGF, cAMP and depolarization.[108,111-116] These observations suggest that it plays a generalized function in regulating cellular responsiveness to external stimuli. Several cis- and trans-acting elements have been identified in the c-fos promoter that are responsible for the complex regulation of c-fos transcription (Fig. 4.7).[117]

One of the first growth factor-regulated elements identified in the c-fos promoter was the serum response element or SRE.[118] The SRE functions as the binding site for the serum response factor (SRF), which is in a complex with a number of other proteins including the ets family transcription regulator p62TCF (also termed Elk1).[119-121] Growth factor treatment results in the rapid phosphorylation of both SRF and p62TCF.[122-124] Phosphorylation of the SRF increases its affinity for the SRE;[122] however, the function of this phosphorylation event in c-fos regulation is unclear, given that the SRF appears to be constitutively bound to the SRE.[125] The activation of c-fos transcription appears to be more tightly linked to the phosphorylation of p62TCF, whose phosphorylation increases its affinity for the c-fos promoter. The kinetics of phosphorylation of p62TCF are correlated with the activation of gene expression.[123,124] p62TCF is likely to be a direct target of the MAP kinases. The MAP kinases phosphorylate p62TCF in vitro, and disruptions of MAP kinase activation also inhibit phosphorylation of p62TCF.[123,124]

A second promoter element, termed the cAMP response element (CRE), mediates the transcriptional activation of c-fos in response to cAMP and to stimuli which elevate intracellular calcium

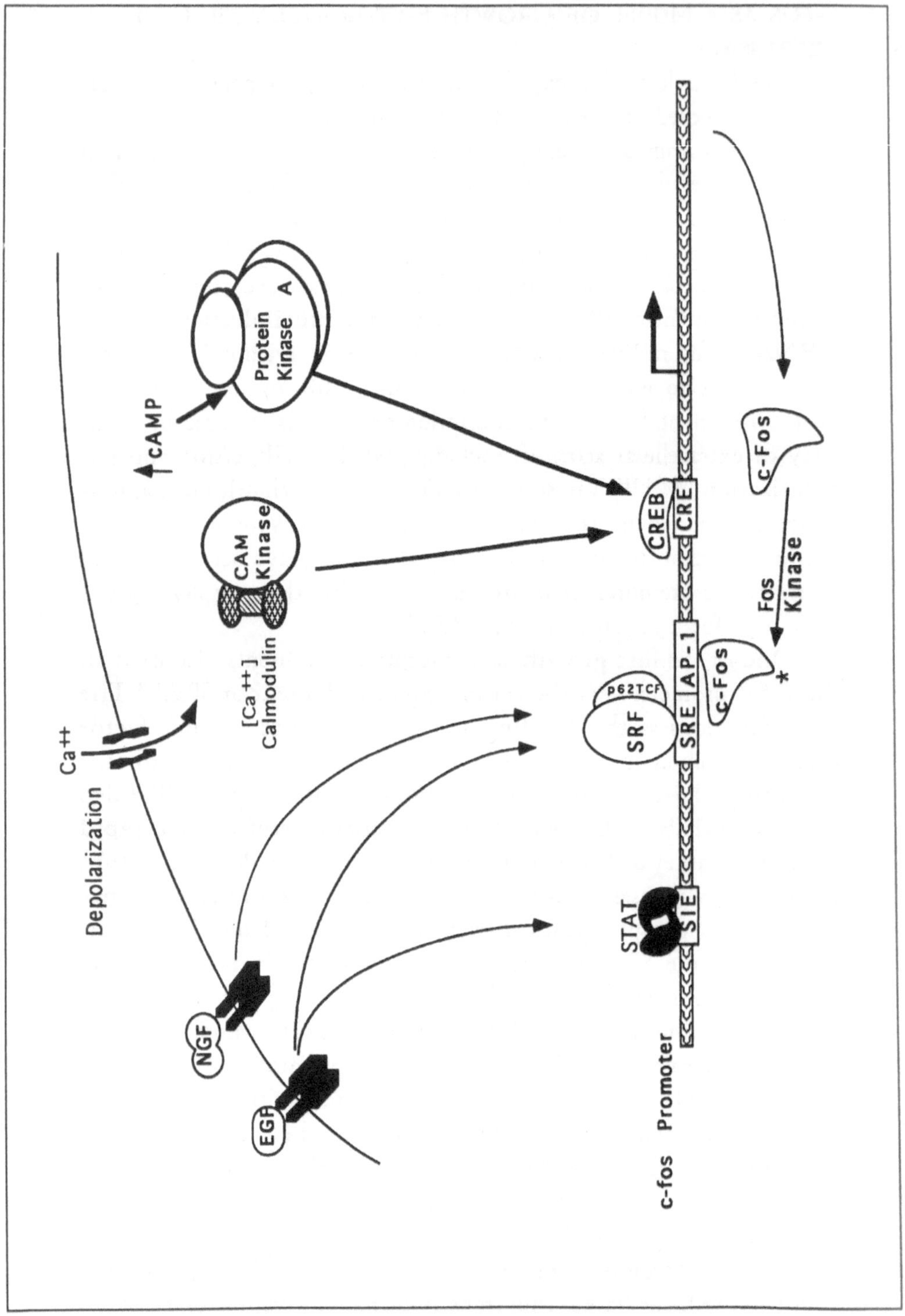
Depolarization
Ca++
[Ca++]
Calmodulin
CAM
Kinase
cAMP
Protein
Kinase A
NGF
EGF
STAT
SIE
SRF
p62TCF
SRE
AP-1
CREB
CRE
c-Fos
Fos
Kinase
c-Fos
*
c-fos Promoter

levels. The protein that mediates this activation, CREB or cAMP response element binding protein, produces its effects by altering protein-protein interactions.[126-128] The ability of CREB to stimulate c-fos transcription is correlated with the phosphorylation of this protein at a unique serine residue (ser133).[129-131] A number of kinases have been identified that can phosphorylate this residue, including PKA, CAM kinase and a novel NGF-induced kinase, termed CREB kinase.[126-128] Phosphorylated CREB species are then capable of participating in the formation of a larger protein complex that assembles on this promoter element in response to growth factor treatment.[129-131]

The last element of the c-fos promoter to be implicated in the activation of c-fos transcription is the SIE or sis-inducible element.[132] This element was originally shown to be responsive to the expression of the sis oncogene or, not surprisingly, to treatments with the sis homolog, PDGF, in some cell types.[132] A similar cis-acting element is also found in a the promoters of a number of interferon-inducible genes.[133] Recently, this element has been shown to function downstream of the JAK family of tyrosine kinases.[134] The JAK kinases have been shown to associate with a number of cytokine and growth factor receptors.[133,134] The targets of the JAK kinases are the SH2 domain-containing transcription factors known as STATs.[134] Phosphorylation of a conserved C-terminal tyrosine results in the dimerization of the STATs through intermolecular interactions between the newly created phosphotyrosine residues and the SH2 domains of their partner.[135,136] Once dimerized, the STATs are rapidly translocated from the membrane to the nucleus where they bind to the SIE and activate transcription.[135,136] This pathway has been shown to be activated in response to a wide range of extracellular ligands, including cytokines and growth factors, such as EGF.[133,134] There are no reports

Fig. 4.7. (on page 82) Regulation of the c-fos promoter. There are three major cis-acting elements within the c-fos promoter that are involved in the extracellular control of c-fos transcription: the sis inducible element (SIE); the serum response element (SRE); and the cAMP/calcium response element (CRE). The SIE is a target of the JAK-STAT pathway and contains a binding site for the STAT transcription factors. The SRE interacts with members of the ELK1/p62TCF family and the serum response factor (SRF). The activity of this element is regulated by phosphorylation of p62TCF and the SRF. The CRE functions through the binding of the CREB protein, whose activity is regulated by a number of kinases that are activated by growth factors, cAMP or calcium.

of this pathway being activated in response to NGF. While the relevance of these findings have not been fully explored in PC12 cells, it suggests that mitogenic signaling may uniquely employ these elements.

The c-fos promoter is subject not only to transcriptional activation, but also to transcriptional repression. Repression of the c-fos promoter is actually mediated by c-fos itself, which acts in an autoinhibitory fashion to arrest transcription of the gene.[137-139] Newly synthesized c-fos protein is hypothesized to bind directly to proteins complexed to its own promoter. A number of studies have shown that phosphorylation of the C-terminus of c-fos is essential for this molecule to exhibit trans-repressive activity.[140,141] Deletion of this domain, as occurs in the oncogenic viral cognate, v-fos, creates proteins that exhibit no trans-represssive activity. Mutations of the C-terminal phosphorylation sites in c-fos have also been shown to abolish its trans-repressive activity.[140] pp90rsk and a novel protein kinase, fos kinase, have been shown to phosphorylate the C-terminus of c-fos in vitro and have been postulated to play a role in the trans-repression of c-fos expression.[140-143] It remains to be seen, however, if these kinases act physiologically to regulate c-fos function in vivo.

The regulation of c-fos transcription illustrates how a complex array of signals are integrated at points where multiple signals converge.[117] Presumably, convergence points, such as the c-fos promoter and other growth factor-responsive promoters, will play critical roles in determining the final cellular response to extracellular stimuli. The decision to proliferate or differentiate is most likely determined by elements that, like the c-fos promoter, act to integrate the signals generated by multiple pathways.

SPECIFICITY OF THE RESPONSE OF PC12 CELLS TO GROWTH FACTORS

One of the enigmatic features of the study of growth factor action is that, despite the specificity and uniqueness of the biological responses of PC12 cells to the various stimuli, the great preponderance of the known intracellular responses to these agents are similar. In PC12 cells, NGF binding to trkA initiates myriad events that eventually alter the pattern of gene expression resulting in PC12 cell differentiation.[144] Many of the signaling events are triggered in response to other differentiating agents such as

FGF or PDGF.[5,145] Paradoxically, treatment of PC12 cells with EGF, which acts as a mitogen,[146] also stimulates these signaling pathways.[5] The entire signaling cascade, from receptor autophosphorylation through immediate early gene induction, is essentially identical whether the cells have been stimulated with mitogens or with agents which provoke their differentiation into a neuronal phenotype.[5] Where differences have been found, they have been subtle and their relevance to the specificity of the response remains unclear.

Many of the differences between signals that differentiate PC12 cells and those that act as mitogens are quantitative. EGF receptors, in contrast to NGF receptors, are rapidly down-regulated by internalization and phosphorylation.[147] The more rapid down regulation of the EGF receptor affects the time course of p21ras activation,[148] which in turn effects the time course of MAP kinase activation.[149] MAP kinase activity in response to EGF returns to basal levels within 60 minutes of stimulation, while MAP kinase activity in response to NGF or FGF remains elevated for several hours.[145,149,150] The sustained activation of the MAP kinases is also correlated with their subcellular localization.[149,151,152] In response to NGF, the MAP kinases are translocated into the nucleus, while in response to EGF, the MAP kinases remain principally in the cytoplasm.[149,151,152]

EGF and insulin treatment do not result in the differentiation of PC12 cells.[5] Overexpression of either the EGF receptor or the insulin receptor, however, changes the specificity of the cellular response.[151,152] Treatment of these cells with EGF or insulin results in the sustained activation and nuclear translocation of the MAP kinases and stimulates the differentiation of these cells.[151,152] It is unclear if these effects are due to the length of time the receptors are activated, the number of activated receptors, or to other non-specific changes in the interactions between the receptors and their targets. Increasing the number of activated receptors, and therefore the concentration of certain phosphotyrosine residues, may result in illegitimate interactions that under normal circumstances would not occur. Similarly, decreasing the number of receptors may block legitimate interactions by lowering the concentration of critical phosphotyrosine residues to levels below that necessary for association with SH2 domain-bearing signal transducing proteins. SH2 domain-bearing proteins exhibit different affinities for individual phosphotyrosine residues.[31] The protein kinase inhibitor, K-252a, also changes the response of PC12 cells to EGF.[153] In the presence of K-252a, EGF treatment differentiates PC12 cells. This change in specificity may be due to the inhibition of specific protein kinases, such as PKC, which phosphorylate and down-regulate the EGF receptor.[147]

These findings have led to the hypothesis that the difference between a differentiative versus a mitogenic signal lies with the strength or duration of the signal.[95] In this model, the subcellular localization of the MAP kinases has been suggested to regulate the specificity of the growth factor response in PC12 cells; signals that elicit MAP kinase nuclear localization result in differentiation, while signals that do not achieve the threshold, either in strength or duration, for MAP kinase translocation to the nucleus result in mitogenesis.[95,149,151,152] The ability of the MAP kinases to directly phosphorylate and alter the behavior of nuclear transcription factors offers an appealing mechanism to explain the change in cellular phenotype that results from MAP kinase nuclear localization. This hypothesis is attractive because it fits well with the idea of morphogenic gradients being important regulators of differentiation. In a number of systems, pluripotent progenitor cells respond to different concentrations of the morphogen by differentiating into distinct cell types, resulting in the formation of unique structures. Morphogenic gradients have been implicated in pattern formation in the embryo, dorsal-ventral polarity, anterior-posterior polarity and in limb development. In most cases, it is unclear how the gradient is formed or how it functions to regulate differentiation or pattern formation. In a model based on strength or duration of the transduced signal, different cells along the morphogenic gradient may achieve threshold levels of signal for differentiation; cells which do not achieve such levels undergo alternative differentiation pathways. However, in this model, it remains unclear how the MAP kinases are directed into the nucleus. The MAP kinases do not contain a canonical nuclear localization signal and nuclear localization occurs independently of their enzymatic activation and appears to be an all or nothing phenomenon.[86,154] It is possible that the localization of the MAP kinases within the nucleus is a result of their retention in the nucleus through interactions with newly synthesized nuclear targets. The strength of the association between the MAP kinase homologue, SAPK, and its substrate, c-jun, provides an example of how this nuclear retention may occur.[97] If such a mechanism operated with the MAP kinases, nuclear retention could be the result of these kinases interacting with a differentiation specific nuclear protein. Moreover, it is not yet proven whether the MAP kinase nuclear localization is the signal for differentiation or is an effect of NGF-stimulated synthesis of proteins required for PC12 cell differentiation.

In addition to these quantitative differences between signals that result in differentiation and those that result in mitogenesis, there are several qualitative differences as well. The most compelling evidence that neuronal differentiation is the result of unique and instructive signals is the discovery of a tyrosine phosphorylated 80 kDa protein termed suc-associated neurotrophic factor-induced tyrosine-phosphorylated target (SNT).[155] SNT is tyrosine-phosphorylated in response to either NGF or FGF, but not in response to EGF.[155] SNT becomes phosphorylated within one minute of NGF treatment and the tyrosine-phosphorylated species is detected only in the nucleus. An intriguing feature of SNT is its ability to bind a protein linked to the control of the cell cycle, suc1. Suc1 directly binds to the cell cycle regulator cdc2 and is used to purify cell cycle regulatory proteins. The ability of SNT to bind to suc1 suggests that SNT may play a role in regulating the cell cycle. Such regulation of the cell cycle may be the key that determines whether PC12 cells divide or differentiate. In response to NGF or FGF, PC12 cells exit the cell cycle and then elaborate neurites.[3] EGF, on the other hand, promotes proliferation rather than differentiation.[146] The presence of potent mitogens, such as serum, results in a delayed response to NGF compared to that measured in the absence of serum. These findings are consistent with the observation that the expression of E1A, a viral protein which promotes progression through the cell cycle, blocks the ability of NGF to differentiate PC12 cells.[156-158] These data indicate that the regulated exit from the cycle may be an obligatory event in the differentiation of PC12 cells.

Currently, there are no other compelling candidates for factors which regulate the specificity of the growth factor response. Clearly, the determinant of specificity must reside with the receptor, either in its length of activation or in its targets. An intriguing possibility is that signals originating from p75 determine the specificity of the response.[14] However, the MAP kinase cascade and cell cycle regulators, which lie downstream of trkA, also play critical roles in determining the specificity of the response.[95] The pace that this field is moving assures that these issues will be laid to rest sooner rather than later.

ACKNOWLEDGMENTS

This work was supported by grants from the National Science Foundation (IBN 10433) and the NIH (NS31987). M.P.M. was supported by a training grant from the NIH (HD 07204-25).

REFERENCES

1. Greene LA, Tischler AS. Establishment of a noradrenergic clonal cell line of rat phaeochromocytoma cells that respond to nerve growth factor. Proc Natl Acad Sci USA 1976; 73:2424-28.
2. Burstein DE, Greene LA. Evidence for RNA synthesis-dependent, and-independent pathways in the stimulation of neurite outgrowth by nerve growth factor. J Cell Biol 1978; 75:6059-63.
3. Greene L, Aletta J, Rukenstein A et al. PC12 pheochromocytoma cells: Culture, NGF treatment and experimental exploitation. Meth Enzymol 1987; 147:207-16.
4. Mesner PW, Winters TR, Green SH. Nerve growth factor withdrawal-induced cell death in neuronal PC12 cells resembles that in sympathetic neurons. J Cell Biol 1992; 119:1669-80.
5. Chao M. Growth factor signaling: where is the specificity? Cell 1992; 68:995-1007.
6. Wood K, Roberts T. Oncogenes and protein kinases in neuronal growth factor action. Biochem Biophys Acta 1993; 1155:133-50.
7. Raffioni S, Bradshaw R, Buxser S. The receptors for nerve growth factor and other neurotrophins. Ann Rev Biochem 1993; 62:823-50.
8. Bothwell M. Keeping track of neurotrophin receptors. Cell 1991; 65:915-18.
9. Chao M. Neurotrophin receptors: A window into neuronal differentiation. Neuron 1992; 9:583-93.
10. Johnson D, Lanahan A, Buck CR et al. Expression and structure of the human NGF receptor. Cell 1986; 47:545-54.
11. Radeke M, Misco T, Hsu C et al. Gene transfer and molecular cloning of the rat nerve growth factor receptor. Nature 1987; 325:593-97.
12. Rodriguez-Tebar A, Dechant G, Barde Y. Binding of brain-derived neurotrophic factor to the nerve growth factor receptor. Neuron 1990; 4:487-92.
13. Rabizadeh S, Oh J, Zhong L et al. Induction of apoptosis by the low affinity NGF receptor. Science 1993; 261:345-48.
14. Dobrowsky RT, Werner MH, Castillino AM et al. Activation of

the sphingomyelin cycle through the low affinity neurotrophin receptor. Science 1994; 265:1596-98.
15. Hempstead B, Martin-Zanca D, Kaplan D et al. High affinity NGF binding requires coexpression of the trk proto-oncogene and the low affinity NGF receptor. Nature 1991; 678-83.
16. Jing S, Tapley P, Barbacid M. Nerve growth factor mediates signal transduction through trk homodimer formation. Neuron 1992; 9:1067-79.
17. Kaplan D, Hempstead B Martin-Zanca D et al. The trk proto-oncogene product: a signal transducing receptor for nerve growth factor. Science 1991; 252:554-58.
18. Kaplan D, Martin-Zanca D, Parada L. Tyrosine phosphorylation and tyrosine kinase activity of the trk proto-oncogene product induced by NGF. Nature 1991; 350:158-60.
19. Klein R, Jing S, Nanduri V et al. The trk proto-oncogene encodes a receptor for NGF. Cell 1991; 65:189-97.
20. Nebreda AR, Martin-Zanca D, Kaplan DR et al. Induction by NGF of meiotic maturation of Xenopus oocytes expressing the trk proto-oncogene product. Science 1991; 252:558-61.
21. Barbacid M, Lamballe F, Pulido D et al. The trk family of tyrosine protein kinase receptors. Biochim Biophys Acta 1991; 1072:115-27.
22. Lee K, Li E, Huber J et al. Targeted mutation of the gene encoding the low affinity NGF receptor p75 leads to deficits in the peripheral sensory nervous system. Cell 1992; 69:737-49.
23. Loeb DM, Maragos J, Martin-Zanca D et al. The trk proto-oncogene rescues NGF responsiveness in mutant NGF-nonresponsive PC12 cell lines. Cell 1991; 66:961-66.
24. Klein R, Nanduri V, Jing S et al. The trkB tyrosine protein kinase is a receptor for brain derived neurotrophic factor and neurotrophin 3. Cell 1991; 66:395-403.
25. Lamballe F, Klein R, Barbacid M. TrkC, a new member of the trk family of tyrosine protein kinases, is a receptor for neurotrophin 3. Cell 1991; 66:967-79.
26. Berg MM, Sternberg DW, Hempsted BL et al. The low-affinity p75 nerve growth factor receptor mediates NGF-induced tyrosine phosphorylation. Proc Natl Acad Sci USA 1991; 88:7106-10.
27. Barker PA, Shooter EM. Disruption of NGF binding to the low affinity neurotrophin receptor p75LNTR reduces NGF binding to trkA on PC12 cells. Neuron 1994; 18:203-15.
28. Hantzopoulos PA, Suri C, Glass DJ et al. The low affinity NGF receptor, p75, can collaborate with each of the trks to potentiate functional responses to the neurotrophins. Neuron 1994; 18:187-201.
29. Rovelli G, Heller RA, Canossa M et al. Chimeric tumor necrosis factor-TrkA receptors reveal that ligand-dependent activation of the

trkA tyrosine kinase is sufficient for differentiation and survival of PC12 cells. Proc Natl Acad Sci USA 1993; 90:8717-21.
30. Koch A, Anderson D, Moran M et al. SH2 and SH3 domains: Elements that control interactions of cytoplasmic signaling proteins. Science 1991; 252:668-74.
31. Songyang Z, Shoelson S, Chaudhuri M et al. SH2 domains recognize specific phosphopeptide sequences. Cell 1993; 72:767-78.
32. Pawson T. Protein modules and signalling networks. Nature 1995; 373:573-80.
33. Soltoff S, Rabin S, Cantley L et al. Nerve growth factor promotes the activation of phosphatidylinositol 3-kinase and its association with the trk tyrosine kinase. J Biol Chem 1992; 267:17472-77.
34. Obermeier A, Halfter H, Wiesmuller K et al. Tyrosine 785 is a major determinant of Trk-substrate interaction. EMBO J 1993; 12:933-41.
35. Obermeier A, Lammers R, Wiesmuller K et al. Identification of trk binding sites for shc and phosphatidylinositol 3'-kinase and formation of a multimeric signaling complex. J Biol Chem 1993; 268:22963-66.
36. Vetter M, Martin-Zanca D, Parada L et al. NGF rapidly stimulates tyrosine phosphorylation of phospholipase C gamma by a kinase activity associated with the product of the trk protooncogene. Proc Natl Acad Sci USA 1991; 88:5650-4.
37. Chung J, Grammer TC, Lemon KP et al. PDGF- and insulin-dependent pp70S6K activation mediated by phosphatidylinositol-3-OH kinase. Nature 1994; 370:71-75.
38. Chung J, Kuo C, Crabtree G et al. Rapamycin-FKBP specifically blocks growth-dependent activation of and signaling by the 70 kd S6 protein kinases. Cell 1992; 69:1227-36.
39. Calvin J, Jonkyeong C, Fiorentino W et al. Rapamycin selectively inhibits interleukin-2 activation of p70 S6 kinase. Nature 1992; 358:70-73.
40. Basu T, Warne PH, Downward J. Role of Shc in the activation of Ras in response to epidermal growth factor and nerve growth factor. Oncogene 1994; 9:3483-91.
41. Ohmichi M, Matuoka K, Takenawa T et al. Growth factors differentially stimulate the phosphorylation of Shc proteins and their association with Grb2 in PC12 pheochromocytoma cells. J Biol Chem 1994; 269:1143-48.
42. Ravichandran KS, Lorenz U, Shoelson SE et al. Interaction of Shc with Grb2 regulates association of Grb2 with mSos. Mol Cell Biol 1995; 15:593-600.
43. Rozakis-Adcock M, McGlade J, Mbamalu G et al. Association of the shc and Grb2/Sem56 Sh2-containing proteins is implicated in activation of the ras pathway by tyrosine kinases. Nature 1992; 360:689-91.

44. Skolnik EY, Lee CH, Batzer A et al. The SH2/SH3 domain-containing protein GRB2 interacts with tyrosine-phosphorylated IRS1 and Shc: implications for insulin control of ras signalling. EMBO J 1993; 12:1929-36.
45. Egan SE, Giddings BW, Brooks MW et al. Association of Sos Ras exchange protein with Grb2 is implicated in tyrosine kinase signal transduction and transformation. Nature 1993; 363:45-51.
46. Buday L, Downward J. Epidermal growth factor regulates p21ras through the formation of a complex of receptor, grb2 adaptor protein and sos nucleotide exchange factor. Cell 1993; 73:611-20.
47. Li N, Daly R, Yajnik V et al. Guanine nucleotide releasing factor hSos1 binds to Grb2 and links receptor tyrosine kinases to Ras signalling. Nature 1993; 363:85-87.
48. Chardin P, Camonis J, Gale N et al. Human Sos1: a guanine nucleotide exchange factor for ras that binds to grb2. Science 1993; 260:1338-43.
49. Boguski M, McCormick F. Proteins regulating ras and its relatives. Nature 1993; 366:643-54.
50. Bollag G, McCormick F. Regulators and effectors of ras proteins. Ann Rev Cell Biol 1991; 7:601-32.
51 Obermeier A, Bradshaw RA, Seedorf K et al. Neuronal differentiation signals are controlled by nerve growth factor receptor/Trk binding sites for SHC and PLC-gamma. EMBO J 1994; 13:1585-90.
52. Blenis J. Signal transduction via the MAP kinases: proceed at your own RSK. Proc Natl Acad Sci USA 1993; 90:5889-92.
53. de Vries-Smits AMM, Burgering BMT, Leevers SJ et al. Involvement of p21ras in activation of extracellular signal-regulated kinase 2. Nature 1992; 357:602-4.
54. Alema S, Casalbore P, Agostini E et al. Differentiation of PC12 pheochromocytoma cells induced by v-src oncogene. Nature 1985; 316:557-9.
55. Kremer N, D'Arcangelo G, Thomas S et al. Signal transduction by NGF and FGF in PC12 cells requires a sequence of src and ras actions. J Cell Biol 1991; 115:809-19.
56. Feig LA, Cooper GM. Inhibition of NIH3T3 cell proliferation by a mutant ras protein with preferential affinity of GDP. Mol Cell Biol 1988; 8:3235-43.
57. Szeberenyi J, Cai H, Cooper G. Effect of a dominant inhibitory Ha-ras mutation on neuronal differentiation of PC12 cells. Mol Cell Biol 1990; 10:5324-32.
58. Bar-Sagi D, Feramisco JR. Microinjection of the ras oncogene protein into PC12 cells induces morphological differentiation. Cell 1985; 42:841-48.
59. Thomas S, DeMarco M, D'Arcangelo G et al. Ras is essential for NGF- and phorbol ester-induced tyrosine phosphorylation of MAP kinases. Cell 1992; 68:1031-40.

60. Wood K, Sarnecki C, Roberts T et al. Ras mediates NGF receptor modulation of three transducing protein kinases: MAP kinase, Raf-1 and RSK. Cell 1992; 68:1041-50.
61. D'Arcangelo G, Halegoua SAA. Branched signaling pathway for nerve growth factor is revealed by src, ras and raf-mediated gene inductions. Mol Cell Biol 1993; 13:3146-55.
62. Crews CM, Alessandrini A, Erikson R. ERKs: Their fifteen minutes has arrived. Cell Growth and Diff 1992; 3:135-42.
63. Davis R. The mitogen activated protein kinase signal transduction pathway. J Biol Chem 1993; 268:14553-56.
64. Van Aelst L, Barr M, Marcus S et al. Complex formation between Ras and Raf and other protein kinases. Proc Natl Acad Sci USA 1993; 90:6213-17.
65. Moodie S, Willumsen B, Weber M et al. Complexes of ras-GTP with raf-1 and mitogen activated protein kinase kinase. Science 1993; 260:1658-61.
66. Vojtek A, Hollenberg S, Cooper J. Mammalian ras interacts directly with the serine/threonine kinase raf. Cell 1993; 74:205-14.
67. Troppmair J, Bruder J, App H et al. Ras controls coupling of growth factor receptors and protein kinase C in the membrane to Raf-1 and B-raf protein serine kinases in the cytosol. Oncogene 1992; 7:1867-73.
68. Koide H, Satoh T, Nakafuku M et al. GTP-dependent association of Raf-1 with Ha-ras: Identification of Raf as a target downstream of Ras in mammalian cells. Proc Natl Acad Sci USA 1993; 90:8683-86.
69. Jaiswal R, Moodie S, Wolfman A et al. The mitogen activated protein kinase cascade is activated by B-raf in NGF-treated PC12 cell through interaction with p21ras. Mol Cell Biol 1994; 14:6944-53.
70. Jelinek T, Catling AD, Reuter CW et al. Ras and Raf-1 form a signaling complex with Mek-1 but not Mek-2. Mol Cell Biol 1994; 14:8212-18.
71. Moodie SA, Paris MJ, Kolch W et al. Association of MEK1 with p21RAS-GMPPNP is dependent on B-Raf. Mol Cell Biol 1994; 14:7153-62.
72. Warne P, Viciana P, Downward J. Direct interaction of ras and the amino-terminal region of raf-1 in vitro. Nature 1993; 364:352-55.
73. Leevers SJ, Paterson HF, Marshall CJ. Requirement for ras in the raf activation is overcome by targeting raf to the plasma membrane. Nature 1994; 369:411-14.
74. Stokoe D, Macdonald SG, Cadwallader K et al. Activation of Raf as a result of recruitment to the plasma membrane. Science 1994; 264:1413-14.
75. Rapp U, Heidecker G, Huleihel M et al. Raf family serine/threonine protein kinases in mitogen signal transduction. Cold Spring

Harbor Symp Quant Biol 1988; 53:173-84.
76. Howe L, Leevers S, Gomez N et al. Activation of the MAP kinase pathway by the protein kinase raf. Cell 1992; 71:335-42.
77. Kyriakis J, App H, Zhang XF et al. Raf-1 activates MAP kinase kinase. Nature 1992; 358:417-21.
78. McDonald S, Crews C, Wu L et al. Reconstitution of the Raf-1-MEK-ERK signal transduction pathway in vitro. Mol Cell Biol 1993; 13:6615-20.
79. Vaillancourt RR, Gardner AM, Johnson GL. B-raf-dependent regulation of the MEK-1/mitogen-activated protein kinase pathway in PC12 cells and regulation by cyclic AMP. Mol Cell Biol 1994; 14:6522-30.
80. Huang W, Alessandrini A, Crews C et al. Raf-1 forms a stable complex with MEK1 and activates MEK1 by serine phosphorylation. Proc Natl Acad Sci USA 1993; 90:10947-51.
81. Ahn N, Seger R, Krebs E. The mitogen activated protein kinase activator. Curr Opin Cell Biol 1992; 4:992-9.
82. Blenis J, Chung J, Erikson E et al. Distinct mechanisms for the activation of the rsk kinases MAP2 kinase pp90rsk, and pp70s6k signaling systems are indicated by inhibition of protein synthesis. Cell Growth and Diff 1991; 2:279-85.
83. Chung J, Pelech S, Blenis J. Mitogen-activated Swiss mouse 3T3 RSK kinases I and II are related to pp44mpk from sea star oocytes and participate in the regulation of pp90rsk activity. Proc Natl Acad Sci USA 1991; 88:4981-5.
84. Grove JR, Price DJ, Banerjee P et al. Regulation of an Epitope-Tagged Recombinant Rsk-1 S6 Kinase by Phorbol Ester and erk/MAP Kinase. Biochemistry 1993; 32:7727-38.
85. Chen R, Sarnecki C, Blenis J. Nuclear localization and regulation of erk and rsk encoded protein kinases. Mol Cell Biol 1992; 12:915-27.
86. Lenormand P, Sardet C, Pages G et al. Growth factors induce nuclear translocation of MAP kinases but not of their activator Map kinase kinase in fibroblasts. J Cell Biol 1993; 122:1079-88.
87. Seth A, Gonzales F, Gupta S et al. Signal transduction within the nucleus by mitogen actived protein kinase. J Biol Chem 1992; 267:24796-806.
88. Oshima M, Sithanandam G, Rapp UR et al. The phosphorylation and activation of B-raf in PC12 cells stimulated by nerve growth factor. J Biol Chem 1991; 266:23753-60.
89. Stephens RM, Sithanandam G, Copeland TD et al. 95-kilodalton B-Raf serine/threonine kinase: identification of the protein and its major autophosphorylation site. Mol Cell Biol 1992; 12:3733-42.
90. Wu J, Dent P, Jelinek T et al. Inhibition of the EGF-activated MAP kinase signaling pathway by adenosine 3',5'-monophosphate. Science 1993; 262:1065-8.

91. Cook SJ, McCormick F. Inhibition by cAMP of Ras-dependent activation of Raf. Science 1993; 262:1069-72.
92. Wartmann M, Davis M. The native structure of the activated raf protein kinase is a membrane-bound multi-subunit complex. J Biol Chem 1994; 269:6695-701.
93. Peraldi P, Frodin M, Barnier JV et al. Regulation of the MAP kinase cascade in PC12 cells: B-Raf activates MEK-1 (MAP kinase or ERK kinase) and is inhibited by cAMP. FEBS Letters 1995; 357:290-6.
94. Young SW, Dickens M, Tavare JM. Differentiation of PC12 cells in response to a cAMP analogue is accompanied by sustained activation of mitogen-activated protein kinase. FEBS Letters 1994; 338:212-6.
95. Marshall CJ. Specificity of receptor tyrosine kinase signaling: Transient versus sustained extracellular signal-regulated kinase activation. Cell 1995; 80:179-85.
96. Cowley S, Paterson H, Kemp P et al. Activation of MAP kinase kinase is necessary and sufficient for PC12 differentiation and for transformation of NIH3T3 cells. Cell 1994; 77:841-52.
97. Kyriakis JM, Banerjee P, Nikolakaki E et al. The stress-activated protein kinase subfamily of c-Jun kinases. Nature 1994; 369:156-60.
98. Lange-Carter C, Pleiman C, Gardner A et al. A divergence in the MAP kinase regulatory network defined by MEK kinase and raf. Science 1993; 260:315-9.
99. Yan M, Dai T, Deak JC et al. Activation of stress-activated protein kinase by MEKK1 phosphorylation of its activator SEK1. Nature 1994; 372:798-800.
100. Sanchez I, Hughes RT, Mayer BJ et al. Role of SAP/ERK kinase-1 in the stress-activated pathway regulating transcription factor c-Jun. Nature 1994; 372:794-8.
101. Lange-Carter CA, Johnson GL. Ras-dependent growth factor regulation of MEK kinase in PC12 cells. Science 1994; 265:1458-61.
102. Rouse J, Cohen P, Trigon S et al. A novel kinase cascade triggered by stress and heat shock that stimulates MAPKAP kinase-2 and phosphorylation of the small heat shock proteins. Cell 1994; 78:1027-37.
103. Derijard B, Raingeaud J, Barrett T et al. Independent human MAP kinase signal transduction pathways defined by MEK and MKK isoforms. Science 1995; 267:682-5.
104. Stokoe D, Campbell DG, Nakielny S et al. MAPKAP kinase-2; a novel protein kinase activated by mitogen-activated protein kinase. EMBO J 1992; 11:3985-94.
105. Shimizu K, Kuroda S, Yamamori B et al. Synergistic activation by RAS and 14-3-3 protein of a mitogen-activated protein kinase kinase kinase. J Biol Chem 1994; 269:22917-20.
106. Curran T, Morgan J. Superinduction of fos by NGF in the pres-

ence of peripherally active benzodiazepines. Science 1985; 229:1265-8.
107. Distel RJ, Speigelman BM. Proto-oncogene c-fos as a transcription factor. Advances in Cancer Research 1990; 55:37-55.
108. Graham R, Gilman M. Distinct protein targets for signals acting at the c-Fos serum response element. Science 1991; 251:189-92.
109. Gizang-Ginsberg E, Ziff E. Nerve growth factor regulates tyrosine hydroxylase gene transcription through a nucleoprotein complex that contains c-fos. Genes Dev 1990; 4:477-91.
110. Lucibello FC, Neuberg M, Hunter JB et al. Transactivation of gene expression by fos protein: Involvement of a binding site for the transcription factor AP-1. Oncogene 1988; 3:43-51.
111. Milbrandt J. Nerve growth factor rapidly induces c-fos in PC12 cells. Proc Natl Acad Sci USA 1986; 83:4789-93.
112. Kruijer W, Schubert D, Verma IM. Induction of the proto-oncogene fos by nerve growth factor. Proc Nat Acad Sci. USA 1985; 82:7330-4.
113. Bartel D, Sheng M, Lau L et al. Growth factors and membrane depolarization activate distinct programs of early response gene expression: dissociation of fos and jun induction. Genes Dev 1989; 3:304-13.
114. Curran T, Morgan J. Barium modulates c-fos expression and post-translational modification. Proc Natl Acad Sci USA 1986; 83:8521-4.
115. Sheng M, Dougan ST, McFadden G et al. Calcium and growth factor pathways of c-fos transcriptional activation require distinct upstream regulatory sequences. Mol Cell Biol 1988; 8: 2787-96.
116. Treisman R. Transient accumulation of c-fos RNA following serum stimulation requires a conserved 5' element and c-fos 3' sequences. Cell 1985; 42:889-902.
117. Robertson LM, Kerppola TK, Vendril M et al. Regulation of c-fos expression in transgenic mice requires multiple interdependent transcription control elements. Neuron 1995; 14:241-52.
118. Treisman R. Identification of a protein-binding site that mediates transcriptional response of the c-fos gene to serum factors. Cell 1986; 46:567-74.
119. Hill CS, Marais R, John S et al. Functional analysis of a growth factor-responsive transcription factor complex. Cell 1993; 73:395-406.
120. Shaw PE, Schroter H, Nordheim A. The ability of a ternary complex to form over the serum response element correlates with serum inducibility of the human c-fos promoter. Cell 1989; 56:563-72.
121. Hipskind R, Rao V, Mueller G et al. Ets-related protein Elk-1 is homologous to the c-Fos regulatory factor p62tcf. Nature 1991; 354:531-4.

122. Rivera VM, Miranti CK, Misra RP et al. A growth factor-induced kinase phosphorylates the serum response factor at a site that regulates its DNA-binding activity. Mol Cell Biol 1993; 13:6260-73.
123. Marais R, Wynne J, Treisman R. The SRF accessory protein Elk-1 contains a growth factor-regulated transcriptional activation domain. Cell 1993; 73:381-93.
124. Gille H, Sharrocks A, Shaw P. Phosphorylation of transcription factor p62tcf by MAP kinase stimulates ternary complex formation at the c-fos promoter. Nature 1992; 358:414-7.
125. Misra V, Rivera V, Wang J et al. The serum response factor is extensively modified by phosphorylation following its synthesis in serum-stimulated fibroblasts. Mol Cell Biol 1991; 11:4545-54.
126. Ginty DD, Bonni A, Greenberg ME. Nerve growth factor activates a Ras-dependent protein kinase that stimulates c-fos transcription via phosphorylation of CREB. Cell 1994; 77:713-25.
127. Sheng M, McFadden G, Greenberg ME. Membrane depolarization and calcium induce c-fos transcription via phosphorylation of transcription factor CREB. Neuron 1990; 4:571-82.
128. Gonzales GA, Montminy MR. Cyclic AMP stimulates somatostatin gene transcription by phosphorylation of CREB at serine 133. Cell 1989; 59:675-80.
129. Kwok RS, Lundblad JR, Chrivia JC et al. Nuclear protein CBP is a coactivator for the transcription factor CREB. Nature 1994; 370:223-6.
130. Chrivia JC, Kwok RP, Lamb N et al. Phosphorylated CREB binds specifically to the nuclear protein CBP. Nature 1993; 365:855-9.
131. Arias J, Alberts AS, Brindle P et al. Activation of cAMP and mitogen responsive genes relies on a common nuclear factor. Nature 1994; 370:226-9.
132. Wagner BJ, Hayes TE, Hoban CJ et al. The SIF binding element confers sis/PDGF inducibility onto the c-fos promoter. EMBO J 1990; 9:4477-84.
133. Darnell JE, Kerr IM, Stark GR. Jak-STAT pathways and transcriptional regulation in response to IFNs and other extracellular signaling proteins. Science 1994; 264:1415-21.
134. Ihle JN, Witthuhn BA, Quelle FW et al. Signaling by the cytokine receptor superfamily: JAKs and STATs. Trends Biol Sci 1994; 19:222-7.
135. Fu XY, Zhang JJ. Transcription factor p91 interacts with the epidermal growth factor receptor and mediates activation of the c-fos gene promoter. Cell 1993; 74:1135-45.
136. Hill CS, Treisman R. Transcriptional regulation by extracellular signals: Mechanisms and specificity. Cell 1995; 80:199-211.
137. Lucibello F, Lowag C, Neuberg M et al. Trans-repression of the mouse c-Fos promoter: A novel mechanism of Fos-mediated transregulation. Cell 1989; 59:999-1007.

138. Sassone-Corsi P, Sisson J, Verma I. Transcriptional autoregulation of the proto-oncogene Fos. Nature 1988; 334:314-9.
139. Wilson T and Treisman R. Fos C-terminal mutations block down-regulation of c-fos transcription following serum stimulation. EMBO J 1988; 7:4193-4202.
140. Ofir R, Dwarki V, Rashid D et al. Phosphorylation of the C terminus of Fos protein is required for transcriptional transrepression of the c-Fos promoter. Nature 1990; 348:80-2.
141. Chen R, Abate C, Blenis J. Phosphorylation of the c-Fos transrepressive domain by mitogen activated protein kinase and 90 kDa ribosomal S6 kinase. Proc Natl Acad Sci USA 1993; 90:10952-56.
142. Taylor LK, Marshak DR, Landreth GE. Identification of a nerve growth factor-and epidermial growth factor-regulated protein kinase that phosphorylates the protooncogene product c-fos. Proc Natl Acad Sci USA 1993; 90:368-372.
143. Nel AE, Taylor LK, Kumar GP et al. Activation of a novel serine/threonine kinase that phosphorylates c-fos upon stimulation of T and B lymphocytes via antigen and cytokine receptors. J Immunology 1993; 152:4347-57.
144. Szeberenyi J, Eberhardt PE. Cellular components of nerve growth factor signaling. Biochim Biophys Acta 1994; 1222:187-202.
145. Heasley LE, Johnson GL. The PDGF receptor induces neuronal differentiation of PC12 cells. Mol Biol Cell 1992; 3:545-53.
146. Huff K, End D, Guroff G. Nerve growth factor-induced alteration in the response of PC12 pheochromocytoma cells to epidermal growth factor. J Cell Biol 1981; 88:189-98.
147. Countaway JL, McQuilkin P, Girones N et al. Multisite phosphorylation of the epidermal growth factor. J Biol Chem 1992; 265:3407-16.
148. Qiu M, Green S. NGF and EGF rapidly activate p21ras in PC12 cells by distinct, convergent pathways involving tyrosine phosphorylation. Neuron 1991; 7:937-46.
149. Nguyen TT, Scimeca JC, Filloux C et al. Co-regulation of the mitogen-activated protein kinase, extracellular signal-regulated kinase 1, and the 90-kDa S6 kinase in PC12 cells. J Biol Chem 1993; 268:9803-10.
150. Traverse S, Gomez N, Paterson H et al. Sustained activation of the mitogen-activated protein (MAP) kinase cascade may be required for differentiation of PC12 cells. Biochem J 1992; 288:351-5.
151. Dikic I, Schlessinger J, Lax I. PC12 cells overexpressing the insulin receptor undergo insulin-dependent neuronal differentiation. Curr Biol 1994; 4:702-8.
152. Traverse S, Seedorf K, Paterson H et al. EGF triggers the neuronal differentiation of PC12 cells that overexpress the EGF receptor. Curr Biol 1994; 4:694-701.

153. Wu CF, Zhang M, Howard BD. K252a potentiates epidermal growth factor-induced differentiation of PC12 cells. J Neurosci 1993; 36:539-50.
154. Gonzalez FA, Seth A, Raden DL et al. Serum-induced translocation of mitogen-activated protein kinase to the cell surface ruffling membrane and the nucleus. J Cell Biol 1993; 122:1089-101.
155. Rabin S, Cleghorn V, Kaplan D. SNT, a differentiation specific target of neurotrophic factor induced tyrosine kinase activity in neurons and PC12 cells. Mol Cell Biol 1993; 13:2203-13.
156. Heasley LE, Benedict S, Gleavey J et al. Requirement of the adenovirus E1A transformation domain 1 for inhibiton of PC12 cell neuronal differentiation. Cell Regulation 1991; 2:479-89.
157. Boulukos KE, Ziff EB. Adenovirus 5 E1A proteins disrupt the neuronal phenotype and growth factor responsiveness of PC12 cells by a conserved region 1-dependent mechanism. Oncogene 1993; 8:237-48.
158. Kalman D, Whittaker K, Bishop JM et al. Domains of E1A that bind p105Rb, p130, and p300 are required to block nerve growth factor-induced neurite growth in PC12 cells. Mol Biol Cell 1993; 4:353-61.

CHAPTER 5

Origin of Adrenal Chromaffin Cells from the Neural Crest

Kristine S. Vogel

Adrenal chromaffin cells, or pheochromocytes, are one of the most experimentally accessible and extensively studied derivatives of the neural crest. Histological techniques based on catecholamine biochemistry allowed early researchers to identify the precursors for adrenal chromaffin cells as they migrate from the primary sympathetic chains into the adrenal primordia. Subsequent production of antisera that recognize catecholamine-synthesizing enzymes allowed further characterization of precursor migration patterns and chromaffin cell differentiation. Recently, identification of numerous molecular markers of neuronal differentiation and transcriptional regulation has led to the development of a model for the sympathoadrenal lineage and responsiveness of these cells to particular environmental cues. I will first describe the migration patterns of the neural crest-derived precursors for adrenal chromaffin cells in avian and mammalian embryos. I will then review the expression of catecholamine-synthesizing enzymes, neuronal markers and transcription factors in the sympathoadrenal lineage, and relate these patterns of differentiation to environmental cues encountered by pheochromocytes.

Genetic Mechanisms in Multiple Endocrine Neoplasia Type 2A, edited by Barry D. Nelkin.

ADRENAL CHROMAFFIN CELLS ARISE FROM THE NEURAL CREST

TRUNK NEURAL CREST CELLS MIGRATE ALONG MEDIAL AND DORSOLATERAL PATHWAYS

The neural crest of vertebrate embryos arises from the lateral edges of the neural plate. In the trunk of the embryo, neural crest cells undergo an epithelial-mesenchymal transition as the neural tube forms. Following this transition, neural crest cells migrate from their dorsal position along two major pathways to give rise to such diverse cell types as sensory and autonomic neurons, Schwann and satellite cells of peripheral nerves and ganglia, adrenal chromaffin cells and melanocytes (Fig. 5.1 A, B).[1, 2]

Vital dyes, radioisotopes, species-specific differences in cell structure and antibodies that recognize epitopes uniquely expressed by neural crest cells have been used to trace the migratory routes of this precursor population. Neural crest cells that migrate along spatially and temporally distinct pathways localize in specific regions and give rise to characteristic cell types. Neural crest cells that migrate ventrally between the neural tube and rostral two-thirds of each somite, following sclerotome cell dispersal, contribute neurons and glia to the sensory dorsal root ganglia and the sympathetic ganglia, as well as chromaffin cells to the adrenal gland (Fig. 5.1 B, C).[3-5] Environmental cues, including extracellular matrix molecules produced by the somites and notochord, influence the precise routes and segmental localization of neural crest cells that migrate on this medial pathway.[6-10] Later in development, a dorsolateral pathway between the dermatomyotome and epidermis becomes available to neural crest cells (Fig. 5.1 C).[11] Cells that migrate on this pathway give rise to melanocytes and perhaps Schwann cells, but never to neurons.[12] In mammalian, avian and zebra fish embryos, it is clear that early-migrating neural crest cells give rise to sensory and sympathetic derivatives, whereas late-migrating neural crest cells give rise to melanocytes.[13-16] Although the phenotypic choices of neural crest cells are correlated with their ultimate destinations, it is not known whether early, premigratory heterogeneity may determine the initial choice of a crest cell for a particular migration pathway, consistent with a *selective* rather than *instructive* role for the environment in neural crest cell differentiation.[17]

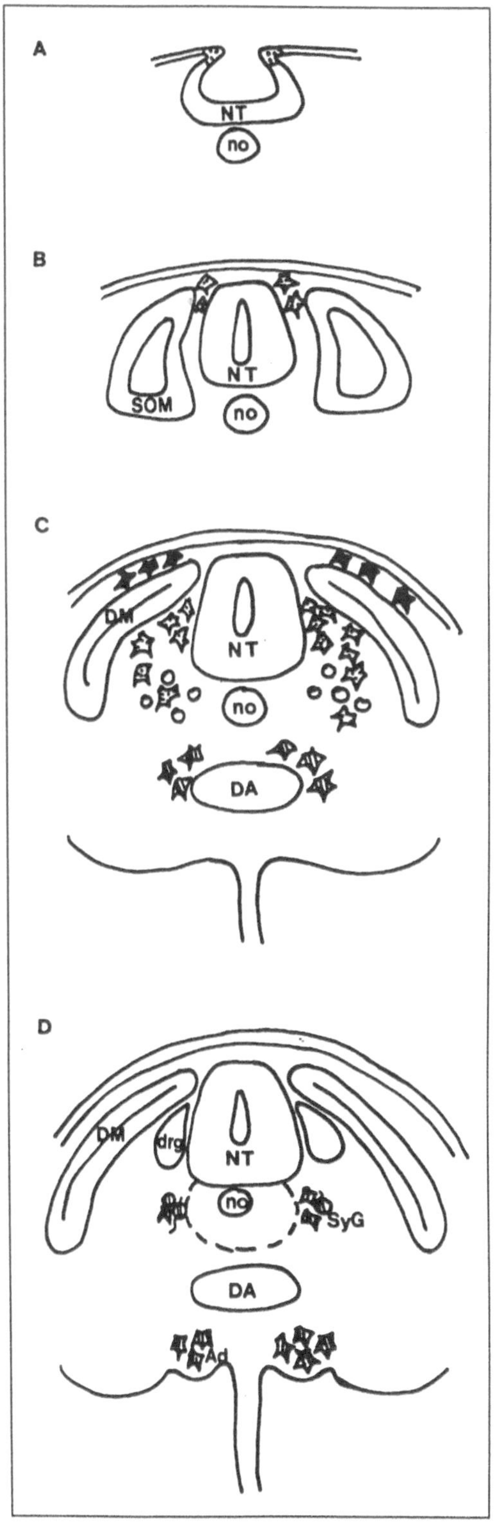

Fig. 5.1. Early migration patterns of neural crest-derived pheochromocyte precursors. Diagrammatic transverse sections through a vertebrate embryo. (A.) The neural crest (stippled) forms from the lateral edges of the neural plate. (B.) Neural crest cells (stippled) undergo an epithelial-mesenchymal transition. (C.) Neural crest cells migrate on ventral (stippled cells) and dorsolateral (solid cells) pathways. The primary sympathetic chains (hatched cells) form near the dorsal aorta. Round open cells represent sclerotome. (D.) Secondary migration of primary chain sympathoblasts into sympathetic ganglia and adrenal primordia (hatched cells). NT-neural tube; no-notochord; SOM-epithelial somite; DM-dermatomyotome of somite; DA-dorsal aorta; DRG-dorsal root ganglion; SyG-sympathetic ganglion; Ad-adrenal gland primordium.

Migration of Neural Crest-Derived Adrenal Chromaffin Cell Precursors

A neural crest origin for sympathetic neurons and adrenal chromaffin cells had long been surmised from the results of histological and extirpation studies in avian and mammalian embryos.[18-21] Weston[22] used tritiated thymidine-labeled grafts in the chicken embryo to confirm that the neural crest is indeed the source of adrenal chromaffin cells. LeDouarin and Teillet[23,24] performed isotopic and isochronic grafts of quail neural primordia into chicken embryos to identify the precise axial level origin of sympathetic neuron and adrenal chromaffin cell precursors. In the avian embryo, neural crest cells that migrate from the neural tube corresponding to somites 8 through 28 contribute catecholaminergic cells to the sympathetic ganglia, aortic and adrenal plexuses and adrenal gland. Specifically, the neural crest region corresponding to somites 18 through 24 gives rise to all of the adrenal chromaffin cells, and is thus termed the "adrenomedullary" level.[2]

The precursors for both sympathetic ganglion neurons and adrenal chromaffin cells initially reside in the primary sympathetic chains, located dorsolateral to the aorta.[18,19,25] Some of these neural crest-derived cells form the definitive sympathetic ganglia of the adult animal, while others migrate ventrally to contribute chromaffin cells to the developing adrenal primordia. In the chicken embryo, the formaldehyde-induced fluorescence (FIF) technique to reveal catecholamines[26] has been used to trace the initial formation of the primary sympathetic chains dorsolateral to the aorta on embryonic day 3.5 (E3.5).[27-29] FIF has also been used in mammalian embryos to follow the subsequent ventral migration of primary chain sympathoblasts into the adrenal gland by E15 in the rat.[30,31] Teitelman et al[32] generated antisera against catecholamine (CA) synthesizing enzymes to follow migration of chromaffin cell precursors from the primary sympathetic chains into the region of the developing suprarenal plexus and subsequently into the adrenal gland primordia. The ontogeny of expression of both catecholaminergic and neuronal markers in adrenal chromaffin cells will be described in more detail in the following section, as a prelude to the identification of a common bipotential precursor for sympathetic neurons and pheochromocytes.

EXPRESSION OF DIFFERENTIATED TRAITS BY ADRENAL CHROMAFFIN CELLS AND THEIR PRECURSORS

CATECHOLAMINE BIOSYNTHETIC ENZYMES ARE EARLY MARKERS FOR CHROMAFFIN CELL PRECURSORS

The primary enzymes involved in catecholamine biosynthesis are listed in Figure 5.2. Several groups of researchers have generated antisera that recognize specific enzymes to study the ontogeny of their appearance in mammalian and avian embryos. Tyrosine hydroxylase (TH), the rate-limiting enzyme that catalyzes conversion of tyrosine to 3,4dihydroxyphenylalanine (dopa), can be detected in sympathoblasts that aggregate dorsolateral to the aorta by E11 in the rat.[32-34] Although these neural crest-derived cells express a differentiated trait characteristic of mature adrenergic neurons, they are still capable of proliferating.[35] TH+ cells populate a plexus near the embryonic kidneys by E14, and begin to invade the defined adrenal primordia by E14.5 in the rat.[32,36] By E16, these TH+ precursors have coalesced within the adrenal cortical tissue to form a medulla.[32] In avian embryos, TH protein is detectable in cells that populate the primary sympathetic chains by E3.5;[37,38] cells that invade the adrenal anlage by E6 continue to express TH. Avian chromaffin cells do not organize into a discrete medulla, and remain interspersed with glucocorticoid-producing cells.[37] Dopamine beta-hydroxylase (DBH), which catalyzes the conversion of dopamine to norepinephrine, appears in primary chain sympathoblasts and adrenal chromaffin cell precursors at the same time as TH.[32,34] Neither TH nor DBH has been detected in migrating neural crest cells prior to their localization in the primary sympathetic chains.[32,34,37,38]

Phenylethanolamine-N-methyl-transferase (PNMT), the enzyme that catalyzes conversion of norepinephrine to epinephrine, is present in 85% of mature adrenal chromaffin cells and is not expressed by sympathetic neurons in the adult rodent.[36,39,40] Originally, PNMT-immunoreactivity was not detected in cells that had invaded the adrenal anlage until E17-E18 in the rat.[32,39,41] However, more recent experiments reveal PNMT protein in pheochromocyte precursors by E15.5-E16.[36,42]

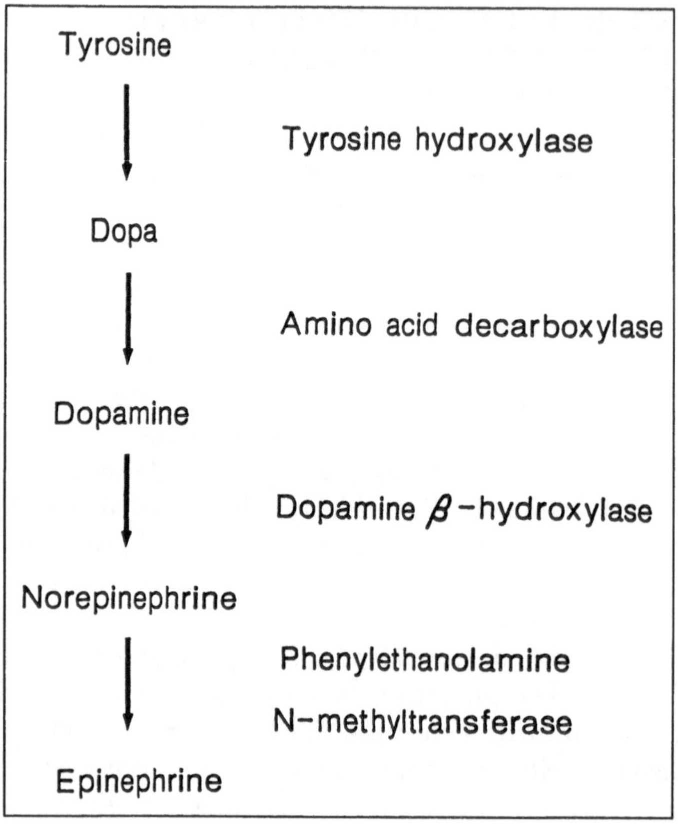

Fig. 5.2. Catecholamine biosynthesis pathway.

Adrenal Chromaffin Cells Lose Neuronal Traits

The precursors for adrenal chromaffin cells express several neuron-specific markers as they migrate from the primary sympathetic chains into the adrenal primordia. In the mouse embryo, sympathoblasts in the primary chains contain all three neurofilament proteins as early as E10.5.[43] SCG10, a membrane-bound protein characteristic of developing central and peripheral neurons,[44] can be detected in primary sympathetic chain cells by E11.5 in the rat.[45,46] Chromaffin cell precursors that enter the developing adrenal gland express SCG10 mRNA transiently; by E14.5, 50% of TH+ adrenal cells continue to express neuronal markers such as the 68-kDa neurofilament protein and SCG10.[25] Expression of SCG10 and neurofilament proteins by TH+ adrenal cells declines throughout embryogenesis and postnatal development, such that in the adult rat, SCG10 mRNA is present in sympathetic neurons but not in adrenal chromaffin cells.[45]

This progressive loss of neuronal traits by adrenal chromaffin cells also occurs in avian embryos. Both SCG10 and the 160 kDa neurofilament protein are present in the primary chain sympathoblasts by E3.5.[37,38] The proportion of TH+ quail adrenal cells that also express the neuron-specific ganglioside recognized by the A2B5 antibody declines throughout embryogenesis, from >95% on E7 to <25% on E15.[37] The loss of neuronal traits by pheochromocyte precursors in vertebrate embryos is thought to occur under the influence of glucocorticoids produced by the adrenal cortical cells, and will be discussed in detail in a later section.

Subpopulations of Adrenal Chromaffin Cells Express Neuropeptides

Neuropeptide Y (NPY) and L-enkephalin (L-Enk) are expressed by subpopulations of adult mammalian pheochromocytes.[47-49] The ontogeny of neuropeptide expression in the rat adrenal gland has been examined in detail by Henion and Landis.[36] NPY+ cells are present in the adrenal primordia by E15; the proportion of NPY+ adrenal cells increases throughout late embryonic development to comprise 80% of TH+ adrenal cells by E20. Postnatally, the proportion of NPY+ pheochromocytes declines to the adult level of 40% by the end of the first week. L-Enk immunoreactivity cannot be detected in TH+ adrenal cells until E16; by E20, 40% of TH+ adrenal cells are also L-Enk+. As for NPY-immunoreactivity, the proportion of L-Enk+ pheochromocytes declines after birth to reach the adult level of 10% by postnatal day 2.[36] In the adult adrenal gland, NPY and L-Enk are restricted to the epinephrine-synthesizing population (PNMT+) of pheochromocytes.[36,50]

Monoclonal Antibodies Identify Chromaffin Cell Precursors

Cyclophosphamide-mediated immunosuppression allowed Carnahan and colleagues to isolate a panel of monoclonal antibodies specific for adult rat chromaffin cells.[46,51,52] All TH+ cells in the primary sympathetic chains express the epitope recognized by the antibody SA1 on E11.5 in the rat. The chromaffin cell precursors that invade the adrenal gland are SA1+ and maintain SA1 immunoreactivity throughout embryonic development.[46] In the adult rat, both PNMT+ and PNMT- chromaffin cells express the SA1 epitope.[51] In contrast, SA1 immunoreactivity is lost in those primary chain precursors that give rise to sympathetic neurons.[46]

Transcription Factors Allow Early Identification of Chromaffin Cell Precursors

Numerous vertebrate homologues of transcription factor genes essential for normal *Drosophila* development have been identified recently, and at least three of these genes appear to have roles in the ontogeny of chromaffin cells. Mammalian achaete-scute homologue-1 (MASH-1), a basic helix-loop-helix transcription factor, is expressed in both embryonic peripheral and central neuronal precursors.[53] Lo et al[54] generated a monoclonal antibody against MASH-1 protein to characterize the expression of this transcription factor in the rat embryo. MASH-1+ cells are present in the primary sympathetic chains dorsolateral to the aorta by E11-E11.5. On E12.5, TH and MASH-1 are expressed in the same regions of sympathetic ganglia, but the strengths of TH and MASH-1 immunoreactivities in individual cells appear to be reciprocal. By E13.5, MASH-1-immunoreactivity can no longer be detected in the sympathetic ganglia, consistent with the proposal that precursor cells down-regulate MASH-1 expression as they increase expression of differentiated catecholaminergic traits.[54] As in the rat embryo, expression of the chicken achaete-scute homologue (CASH-1) by the primary chain sympathoblasts is extinguished as these cells increase expression of TH. The MASH-1 gene has been targeted by homologous recombination in the mouse; although homozygous mutant embryos lack sympathetic neurons and parasympathetic ganglia, only a partial reduction in the number of adrenal chromaffin cells was observed.[55]

The homeodomain-containing vertebrate paired-like gene, Phox2, is expressed in sympathetic, parasympathetic and enteric neurons and in the adrenal medulla by E13.5 in the mouse. The onset of Phox2 expression coincides with production of TH and DBH in the primary sympathetic chains, and is thought to be part of a regulatory cascade that initiates the adrenergic phenotype.[56] In the chick, the onset of Phox2 expression closely follows that of CASH-1 in the primary sympathetic chains.[38] Expression of GATA-2, a zinc-finger transcription factor gene, coincides with the appearance of Phox2 in avian sympathoblasts.[38]

ENVIRONMENTAL CUES INFLUENCE THE FATES OF SYMPATHOADRENAL PRECURSORS

The origin of both sympathetic neurons and pheochromocytes from the primary sympathetic chains, the common expression of neuronal traits and transcription factors and the ability of pheochromocytes from both fetal and neonatal mammals to acquire neuronal traits[57-59] are consistent with the existence of a common bipotential precursor for these two cell types. As described above, this neural crest-derived precursor for the sympathoadrenal lineage initially expresses neuronal traits.[25,37,45] Precursor cells that migrate into the definitive sympathetic ganglia continue to express neuronal characters and become dependent on nerve growth factor (NGF) for survival. In contrast, those precursors that migrate into the adrenal gland lose neuronal traits under the influence of glucocorticoids. The responsiveness of sympathoadrenal precursors to identified environmental cues throughout development has been well-documented in both avian and mammalian embryos, and will be reviewed in the following sections.

TISSUE INTERACTIONS INFLUENCE THE CATECHOLAMINERGIC DIFFERENTIATION OF NEURAL CREST CELLS

Both in vitro and in vivo studies with avian neural crest cells are consistent with an essential role for environmental cues encountered during migration in the acquisition of the catecholaminergic phenotype. These cues are thought to be produced by cells in the ventral neural tube, notochord and somitic mesenchyme. Cohen[60] used cultures of neural crest cells on the chick chorioallantoic membrane to demonstrate that interactions with those cell populations encountered during migration were required for catecholaminergic differentiation. Norr[61] examined interactions between neural crest cells and neighboring tissues in organ cultures to conclude that the neural tube and notochord induce changes in somitic mesenchyme, which in turn influences catecholaminergic differentiation in neural crest cells. Neural tube-conditioned medium can promote the development of adrenergic neuroblasts in cultures of quail neural crest cells.[62]

Notochord and/or neural tube ablations in the chick embryo also indicate that these tissues have a direct effect on the differentiation of the sympathoadrenal derivatives.[63] Stern and colleagues[64] rotated the neural tube, with or without the notochord, to show that either the notochord in its normal ventral location or the presence of floor plate/ventral motor column region of the neural tube is sufficient to allow catecholaminergic cells to differentiate near the aorta. Recently, Groves and colleagues[38] have demonstrated that the appearance of transcription factor markers Phox2 and GATA-2, as well as of the sympathoadrenal markers TH and catecholamine fluorescence, is dependent on signals derived from the neural tube. In contrast, the expression of the CASH-1 transcription factor gene and the neuronal marker SCG-10 by neural crest-derived cells does not require the presence of the neural tube.

Cultures of avian neural crest cells have been used in an attempt to identify molecules involved in the induction of the catecholaminergic phenotype. Extracellular matrix molecules present in preparations of basement membrane promote catecholaminergic differentiation in neural crest cells,[65] as does the purified matrix molecule, fibronectin.[66,67] Neural crest cells encounter these extracellular matrix components during their ventral migration to the primary sympathetic chains.[10, 68] Transforming growth factor β1 (TGF-β1) , which can enhance production of extracellular matrix,[69] supports catecholaminergic differentiation of quail neural crest cells in vitro.[70] This induction differs from that promoted by chicken embryo extract; neural crest cells can be induced to become CA+ only when exposed to embryo extract within the first four days of culture,[71] whereas TGF-β1 can promote the development of CA+ neural crest cells as late as seven days in vitro.[70] The factors present in chicken embryo extract that promote catecholaminergic differentiation have not been identified. In addition to their effects on pheochromocyte precursors, glucocorticoids can stimulate adrenergic differentiation of migrating and premigratory neural crest cell populations.[72]

Glucocorticoids Repress Neuronal Differentiation and Promote Expression of Chromaffin Cell Traits in Sympathoadrenal Precursors

Neural crest-derived sympathoadrenal precursors that migrate from the primary sympathetic chains into the adrenal primordia are exposed to glucocorticoids (GC) produced by the mesoderm-derived cortical cells. Unsicker and colleagues [57,73] showed that the synthetic GC, dexamethasone, can inhibit NGF-induced neuronal differentiation of embryonic and neonatal adrenal chromaffin cells.

In the absence of NGF or GC, 80% of TH+ adrenal cells isolated from E14.5 rats express neurofilament proteins and extend neurites in vitro; in the presence of dexamethasone, these TH+ cells extinguish neuronal traits.[25] Sympathoadrenal precursors isolated from E14.5 rat adrenals using the monoclonal antibody SA1 retain a rounded morphology, fail to maintain neurofilament expression or to extend neurites and continue to express SA1 epitope when cultured in the presence of dexamethasone.[52] Glucocorticoids can repress neuronal traits in cultured neural crest cells as well. Thus, dexamethasone treatment decreases the proportion of TH+ quail neural crest cells that also express a neuron-specific glycolipid and display a neuronal morphology.[74]

In addition to their negative effects on neuronal differentiation, glucocorticoids promote the survival of adrenal chromaffin cells and regulate the expression of PNMT. The promoter for the rat PNMT gene contains a functional consensus GC-response element.[75] Dexamethasone enhances the survival of embryonic rat chromaffin precursors in vitro[25,59,76] and allows maintenance of PNMT, adrenalin synthesis and large CA storage vesicles in chromaffin cells isolated from young rats.[59] Although glucocorticoids are required to maintain PNMT catalytic activity in fetal and neonatal rat pheochromocytes,[77-79] the precocious expression of PNMT during embryogenesis cannot be forced by exposure to these steroid hormones. Thus, Teitelman et al[32] demonstrated that the appearance of PNMT in embryonic chromaffin cells is not accelerated by treatment of pregnant rats with dexamethasone. Bohn and colleagues[39] extended these observations in a detailed in vivo analysis of GC effects on PNMT expression during rat embryogenesis. Treatment of pregnant mothers or embryos with GC, adrenocorticotrophic hormone, or agents that stabilize PNMT failed to elicit precocious appearance of the enzyme in embryonic pheochromocytes. Conversely, the onset of PNMT expression was not inhibited by embryonic hypophysectomy, steroid antagonists, or GC receptor agonists.[39] Thus, initial competence of chromaffin cells to produce PNMT is not dependent upon glucocorticoids. After the initial appearance of PNMT, however, the levels of this enzyme in cultured adrenal chromaffin cells must be regulated and maintained by GC.[80,81]

Michelson and Anderson[82] performed an elegant series of experiments to identify the molecular mechanisms that underlie the two effects of GC on embryonic adrenal chromaffin cells. Dexamethasone inhibits neurite outgrowth in sympathoadrenal precursors isolated from E14.5 rats as early as 15 hours in vitro; the appearance of PNMT parallels that in vivo and occurs after one to two days in culture. Both effects are mediated by the type II GC receptor; however, a lower concentration of GC is required to inhibit neurite outgrowth than to induce PNMT. Although GC are necessary to induce expression of PNMT in cultured sympathoadrenal precursors, the timing of PNMT induction depends on acquisition of competence that is controlled by a GC-independent cellular "clock."[82]

NGF and FGF Promote Neuronal Differentiation of Sympathoadrenal Precursors and Adrenal Chromaffin Cells

NGF, a survival factor for sympathetic neurons, can induce neuronal differentiation of mammalian adrenal chromaffin cells in vitro and in vivo. Unsicker et al[57] demonstrated that in adrenal chromaffin cells isolated from 7 to 12-day-old rats, NGF promotes neurite outgrowth. Aloe and Levi-Montalcini[58] administered NGF to rat fetuses and pups and observed a transformation of pheochromocytes into sympathetic neurons that extended neurites into the adrenal cortex. Neonatal rat chromaffin cells grown in the presence of NGF for several weeks acquire many traits characteristic of mature sympathetic neurons, including specialized synapses, tetanus toxin labeling and neurofilament protein. These cells also lose chromaffin granules and PNMT.[59] In contrast, NGF does not increase the proportion of TH+ sympathoadrenal precursors isolated from E14.5 rats[25] or in cultures of quail neural crest cells.[74] NGF promotes the proliferation of chromaffin cells isolated from eight day-old rats, and those cells that divide in response to NGF can subsequently acquire neuronal traits.[83] Although adult bovine pheochromocytes do not extend neurites in the presence of NGF,[84] adult rat pheochromocytes do respond to NGF with proliferation and neuronal differentiation.[85]

Fibroblast growth factor, which exists in acidic (aFGF) and basic (bFGF) forms, may act prior to NGF in the embryo to promote proliferation and neuronal differentiation of sympathoadrenal precursors. Basic FGF acts as a mitogen for neonatal rat chromaffin cells and promotes neurite outgrowth and SCG-10 expression

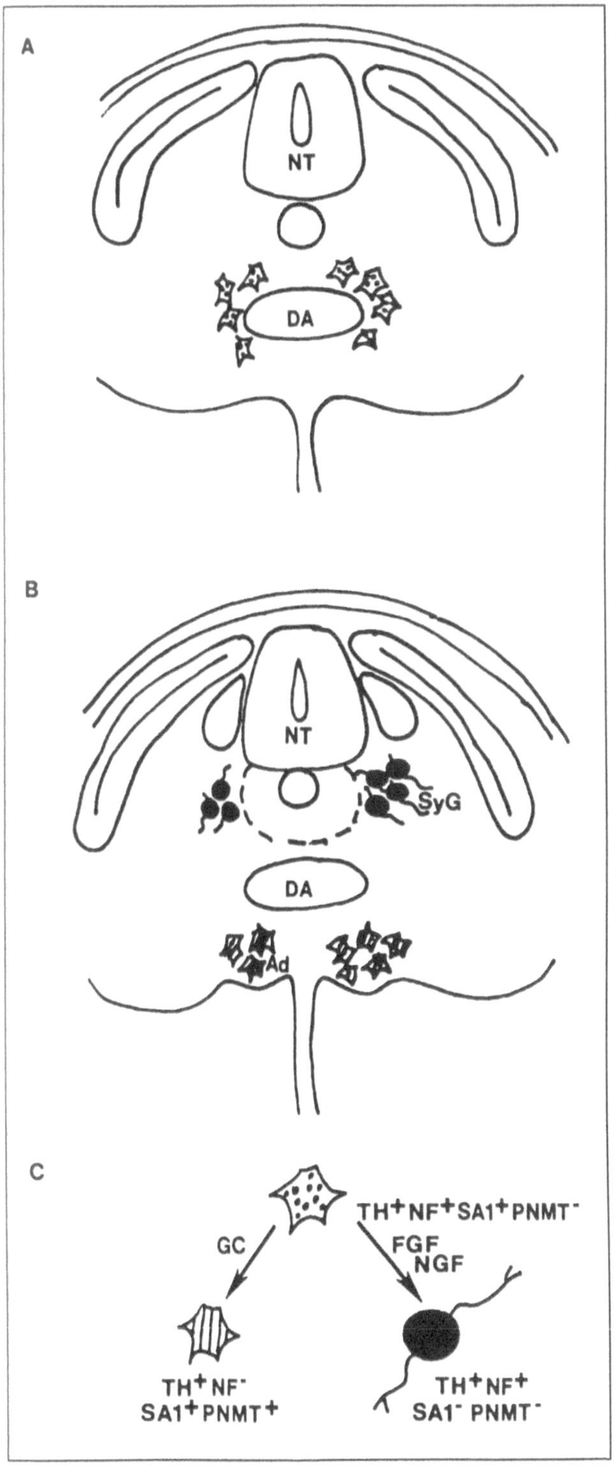

Fig. 5.3 Neural crest-derived sympathoadrenal precursors give rise to sympathetic neurons and adrenal chromaffin cells. Diagrammatic transverse sections through a vertebrate embryo. (A.) Sympathoadrenal precursors initially reside in the primary sympathetic chains near the dorsal aorta (stippled cells). (B.) Sympathoadrenal precursors undergo secondary migration to give rise to sympathetic neurons (solid cells) in the sympathetic ganglia (SyG) and to adrenal chromaffin cells (hatched cells) in the adrenal primordium (Ad). (C.) Environmental cues influence a common bipotential precursor (stippled cell) to differentiate as adrenal chromaffin cells (hatched cell) and sympathetic neurons (solid cell). TH-tyrosine hydroxylase; NF-neurofilament protein; SA1-sympathoadrenal antigen; PNMT-phenylethanolamine N-methyltransferase; GC-glucocorticoids; FGF-figroblast growth factor; NGF-nerve growth factor.

in vitro. Although bFGF cannot support long-term survival of chromaffin cells, it does induce dependence on NGF.[86] Moreover, bFGF acts synergistically with insulin-like growth factors to promote proliferation and survival of neonatal adrenal chromaffin cells.[87] Acidic FGF also stimulates proliferation and neurite outgrowth in young rat pheochromocytes; the action of aFGF on neurite outgrowth can be potentiated by the glycosaminoglycan heparin.[88]

Based on many of the results presented in this section, Anderson[89] has proposed a model for the development of sympathetic neurons and adrenal chromaffin cells from a neural crest-derived bipotential precursor (Fig. 5.3). Under the influence of signals produced by the ventral neural tube and notochord, neural crest cells that localize near the dorsal aorta acquire catecholaminergic traits and express neuronal characteristics. FGF may act early to expand the population of sympathoadrenal precursors; those cells that localize in the region of the developing sympathetic ganglia continue to express neuronal traits and acquire dependence on NGF for long-term survival. Sympathoadrenal precursors that migrate into the adrenal anlage are exposed to GC and begin to lose neuronal characteristics. These cells become competent to transcribe the PNMT gene, and the levels of this enzyme are increased and maintained by glucocorticoids. In conclusion, the sympathoadrenal lineage has provided an experimentally accessible developmental system in which to monitor in detail the consequences of identified cell interactions and molecular cues on phenotypic choices during embryogenesis.

References

1. Weston JA. The migration and differentiation of neural crest cells. Adv Morphogen 1970; 8:41-114.
2. LeDouarin NM. The neural crest. Cambridge: Cambridge University Press, 1982.
3. Rickmann M, Fawcett JW, Keynes RJ. The migration of neural crest cells and the growth of motor axons through the rostral half of the chick somite. J Exp Morphol Embryol 1985; 90:437-55.
4. Bronner-Fraser M. Analysis of the early stages of trunk neural crest migration in avian embryos using the monoclonal antibody HNK-1. Dev Biol 1986; 115:44-55.
5. Loring J, Erickson C. Neural crest cell migration pathways in the chick embryo. Dev Biol 1987; 121:230-6.

6. Teillet MA, Kalcheim C, LeDouarin NM. Formation of the dorsal root ganglion in the avian embryo: segmental origin and migratory behavior of neural crest progenitor cells. Dev Biol 1987; 120:329-47.
7. Lallier T, Bronner-Fraser M. A spatial and temporal analysis of dorsal root and sympathetic ganglion formation in the avian embryo. Dev Biol 1988; 127:99-112.
8. Kalcheim C, Teillet MA. Consequences of somite manipulation on the pattern of dorsal root ganglion development. Development 1989; 106:85-93.
9. Goldstein RS, Kalcheim C. Normal segmentation and size of the primary sympathetic ganglia depend upon the alternation of rostrocaudal properties of the somites. Development 1991; 112:327-34.
10. Tosney KW, Dehnbostel DB, Erickson CA. Neural crest cells prefer the myotome's basal lamina over the sclerotome as a substratum. Dev Biol 1994; 163:389-406.
11. Erickson CA, Duong TD, Tosney KW. Descriptive and experimental analysis of the dispersion of neural crest cells along the dorsolateral path and their entry into ectoderm in the chick embryo. Dev Biol 1992; 151:251-72.
12. Weston JA. Sequential segregation and fate of developmentally restricted intermediate cell populations in the neural crest lineage. Curr Topics Dev Biol 1991; 23:133-53.
13. Vogel KS, Weston JA. A subpopulation of cultured avian neural crest cells has transient neurogenic potential. Neuron 1988; 1:569-77.
14. Artinger KB, Bronner-Fraser M. Partial restriction in the developmental potential of late emigrating avian neural crest cells. Dev Biol 1992; 149:149-57.
15. Raible DW, Eisen JS. Restriction of neural crest cell fate in the trunk of the embryonic zebrafish. Development 1994; 120:495-503.
16. Serbedzija GN, Bronner-Fraser M, Fraser SE. Developmental potential of trunk neural crest cells in the mouse. Development 1994; 120:1709-18.
17. Erickson CA, Goins TL. Avian neural crest cells can migrate in the dorsolateral path only if they are specified as melanocytes. Development 1995; 121:915-24.
18. Rau AS, Johnston PH. Observations on the development of the sympathetic system and suprarenal bodies in the sparrow. Proc Zool Soc London 1923; 3:741-68.
19. Willier BH. A study of the origin and differentiation of the suprarenal gland in the chick embryo by chorio-allantoic grafting. Physiol Zool 1930; 3:201-25.
20. Pankratz DS. The development of the suprarenal gland in the albino rat. Anat Rec 1931; 49:31-9.

21. Hammond WS, Yntema CL. Depletions in the thoraco-lumbar sympathetic system following removal of neural crest in the chick. J Comp Neurol 1947; 86:237-65.
22. Weston JA. A radioautographic analysis of the migration and localization of trunk neural crest cells in the chick. Dev Biol 1963; 6:279-310.
23. LeDouarin NM, Teillet MA. Localisation, par la methode des greffes interspecifiques, du territoire neural dont derivent les cellules adrenales surrenaliennes chez l'embryon d'Oiseau. CR Acad Sci 1971; 272:481-4.
24. Teillet MA, LeDouarin NM. Determination par la methode des greffes heterospecifiques d'ebauches neurales de Caille sur l'embryon de polet, du niveau du nevraxe dont derivent les cellules medullo-surrenaliennes. Arch Anat Microsc Morphol Exp 1974; 63:51-62.
25. Anderson DJ, Axel R. A bipotential neuroendocrine precursor whose choice of cell fate is determined by NGF and glucocorticoids. Cell 1986; 47:1079-90.
26. Falck B. Observations on the possibilities of the cellular localization of monoamines by a fluorescence method. Acta Physiol Scand 1962; Suppl 197.
27. Enemar A, Falck B, Hakanson R. Observations on the appearance of norepinephrine in the sympathetic nervous system of the chick embryo. Dev Biol 1965; 11:268-83.
28. Kirby ML, Gilmore SA. A correlative histofluorescence and light microscopic study of the formation of the sympathetic trunks in chick embryos. Anat Rec 1976; 186:437-50.
29. Allan IJ, Newgreen DF. Catecholamine accumulation in neural crest cells and the primary sympathetic chain. Amer J Anat 1977; 149:413-21.
30. Fernholm M. On the development of the sympathetic chain and adrenal medulla in the mouse. Z Anat Entwicklungsgesch 1971; 133:305-17.
31. Polak JM, Rost FWD, Pearse AGE. Fluorogenic amine tracing of neural crest derivatives forming the adrenal medulla. Gen Comp Endocrinol 1971; 16:132-6.
32. Teitelman G, Baker H, Joh TH et al. Appearance of catecholamine-synthesizing enzymes during development of the rat sympathetic nervous system: Possible role of tissue environment. Proc Natl Acad Sci USA 1979; 76:509-13.
33. Cochard P, Goldstein M, Black IB. Ontogenic appearance and disappearance of tyrosine hydroxylase and catecholamines in the rat embryo. Proc Natl Acad Sci USA 1978; 75:2986-90.
34. Cochard P, Goldstein M, Black IB. Initial development of the noradrenergic phenotype in autonomic neuroblasts of the rat embryo in vivo. Dev Biol 1979; 71:100-14.
35. Rothman TP, Gershon MD, Holtzer H. The relationship of cell

division to the acquisition of adrenergic characteristics by developing sympathetic ganglion cell precursors. Dev Biol 1978; 65:322-41.
36. Henion PD, Landis SC. Asynchronous appearance and topographic segregation of neuropeptide-containing cells in the developing rat adrenal medulla. J Neurosci 1990; 10: 2886-96.
37. Vogel KS, Weston JA. The sympathoadrenal lineage in avian embryos 1. Adrenal chromaffin cells lose neuronal traits during embryogenesis. Dev Biol 1990; 139:1-12.
38. Groves AK, George KM, Tissier-Seta JP et al. Differential regulation of transcription factor gene expression and phenotypic markers in developing sympathetic neurons. Development 1995; 121:887-901.
39. Bohn MC, Goldstein M, Black IB. Role of glucocorticoids in expression of the adrenergic phenotype in rat embryonic adrenal gland. Dev Biol 1981; 82:1-10.
40. Bohn MC, Goldstein M, Black IB. Expression of phenylethanolamine N-methyltransferase in rat sympathetic ganglia and extra-adrenal chromaffin tissue. Dev Biol 1982; 89:299-308.
41. Verhofstad AAJ, Hokfelt T, Goldstein M et al. Appearance of tyrosine hydroxylase, aromatic amino-acid decarboxylase, dopamine B-hrdroxylase and phenylethanolamine N-methyltransferase during the ontogenesis of the adrenal medulla: An immunohistochemical study in the rat. Cell Tissue Res 1979; 200:1-13.
42. Ehrlich ME, Evinger M, Regunathan S et al. Mammalian adrenal chromaffin cells coexpress the epinephrine-synthesizing enzyme and neuronal properties in vivo and in vitro. Dev Biol 1994; 163:480-90.
43. Cochard P, Paulin D. Initial expression of neurofilaments and vimentin in the central and peripheral nervous system of the mouse embryo in vivo. J Neurosci 1984; 4:2080-94.
44. Stein R, Mori N, Matthews K. The NGF-inducible SCG-10 mRNA encodes a novel membrane-bound protein present in growth cones and abundant in developing neurons. Neuron 1988; 1:463-76.
45. Anderson DJ, Axel R. Molecular probes for the development and plasticity of neural crest derivatives. Cell 1985; 42:649-62.
46. Anderson DJ, Carnahan JF, Michelson A et al. Antibody markers identify a common progenitor to sympathetic neurons and chromaffin cells in vivo and reveal the timing of commitment to neuronal differentiation in the sympathoadrenal lineage. J Neurosci 1991; 11:3507-19.
47. Schultzberg M, Lundberg JM, Hokfelt T et al. Enkephalin-like immunoreactivity in gland cells and nerve terminals of the adrenal medulla. Neurosci 1978; 3:1169-86.
48. Bohn MC, Kessler JA, Golightly L et al. Appearance of enkephalin-immunoreactivity in rat adrenal medulla following treatment with nicotinic antagonists or reserpine. Cell Tissue Res 1983;

231:469-79.
49. deQuidt ME, Emson PC. Neuropeptide Y in the adrenal gland: characterization, distribution and drug effects. Neurosci 1986; 19:1011-22.
50. Livett BG, Day R, Elde RP et al. Co-storage of enkephalins and adrenaline in the bovine adrenal medulla. Neurosci 1982; 7:1323-32.
51. Carnahan JF, Patterson PH. The generation of monoclonal antibodies that bind preferentially to adrenal chromaffin cells and the cells of embryonic sympathetic ganglia. J Neurosci 1991; 11:3493-506.
52. Carnahan JF, Patterson PH. Isolation of the progenitor cells of the sympathoadrenal lineage from embryonic sympathetic ganglia with the SA monoclonal antibodies. J Neurosci 1991; 11:3520-30.
53. Johnson JE, Birren SJ, Anderson DJ. Two rat homologues of Drosophila achaete-scute specifically expressed in neuronal precursors. Nature 1990; 346:858-61.
54. Lo LC, Johnson JE, Wuenschell CW et al. Mammalian achaete-scute homolog-1 is transiently expressed by spatially restricted subsets of early neuroepithelial and neural crest cells. Genes Dev 1991; 5:1524-37.
55. Guillemot F, Lo LC, Johnson JE et al. Mammalian achaete-scute homolog 1 is required for the early development of olfactory and autonomic neurons. Cell 1993; 75:463-76.
56. Valarche I, Tissier-Seta JP, Hirsch MR et al. The mouse homeodomain protein Phox2 regulates Ncam promoter activity in concert with Cux/CDP and is a putative determinant of neurotransmitter phenotype. Development 1993; 119:881-96.
57. Unsicker K, Krisch B, Otten U et al. Nerve growth factor-induced fiber outgrowth from isolated rat adrenal chromaffin cells: impairment by glucocorticoids. Proc Natl Acad Sci USA 1978; 75:3498-3502.
58. Aloe L, Levi-Montalcini R. Nerve growth factor-induced transformation of immature chromaffin cells in vivo into sympathetic neurons: effect of antiserum to nerve growth factor. Proc Natl Acad Sci USA 1979; 76:1246-50.
59. Doupe AJ, Landis SC, Patterson PH. Environmental influences in the development of neural crest derivatives: glucocorticoids, growth factors, and chromaffin cell plasticity. J Neurosci 1985; 5:2119-42.
60. Cohen AM. Factors directing the expression of sympathetic nerve traits in cells of neural crest origin. J Exp Zool 1972; 179:167-82.
61. Norr SC. In vitro analysis of sympathetic neuron differentiation from chick neural crest cells. Dev Biol 1973; 34:16-38.
62. Howard MJ, Bronner-Fraser M. The influence of neural tube-derived factors on differentiation of neural crest cells in vitro 1. Histochemical study on the appearance of adrenergic cells. J Neurosci

1985; 5:3302-9.
63. Teillet MA, LeDouarin NM. Consequences of neural tube and notochord excision on the development of the peripheral nervous system in the chick embryo. Dev Biol 1983; 98:192-211.
64. Stern CD, Artinger KB, Bronner-Fraser M. Tissue interactions affecting the migration and differentiation of neural crest cells in the chick embryo. Development 1991; 113:207-16.
65. Maxwell GD, Forbes ME. Exogenous basement membrane-like matrix stimulates adrenergic development in avian neural crest cultures. Development 1987; 101:767-76.
66. Sieber-Blum M, Sieber F, Yamada KM. Cellular fibronectin promotes adrenergic differentiation of quail neural crest cells in vitro. Exp Cell Res 1981; 133:285-95.
67. Loring J, Glimelius B, Weston JA. Extracellular matrix materials influence quail neural crest cell differentiation in vitro. Dev Biol 1982; 90:165-74.
68. Newgreen DF, Thiery JP. Fibronectin in early avian embryos: synthesis and distribution along the migration pathways of neural crest cells. Cell Tissue Res 1980; 211:269-91.
69. Rogers SL, Gegick PJ, Alexander SM et al. Transforming growth factor-B alters differentiation in cultures of avian neural crest-derived cells: Effects on cell morphology, proliferation, fibronectin expression, and melanogenesis. Dev Biol 1992; 151:192-203.
70. Howard MJ, Gershon MD. Role of growth factors in catecholaminergic expression by neural crest cells: in vitro effects of transforming growth factor beta-1. Dev Dynamics 1993; 196:1-10.
71. Maxwell GD, Forbes ME. Stimulation of adrenergic development in neural crest cultures by a reconstituted basement membrane-like matrix is inhibited by agents that elevate cAMP. J Neurosci Res 1990; 25:172-9.
72. Smith J, Fauquet M. Glucocorticoids stimulate adrenergic differentiation in cultures of migrating and premigratory neural crest. J Neurosci 1984; 4:2160-72.
73. Unsicker K, Millar TJ, Muller TH et al. Embryonic rat adrenal glands in organ culture: Effects of dexamethasone, nerve growth fator and its antibodies on pheochromoblast differentiation. Cell tissue Res 1985; 241:207-17.
74. Vogel KS, Weston JA. The sympathoadrenal lineage in avian embryos II. Effects of glucocorticoids on cultured neural crest cells. Dev Biol 1990; 139:13-23.
75. Ross ME, Evinger MJ, Hyman SE et al. Identification of a functional glucocorticoid response element in the phenylethanolamine N-methyltransferase promoter using fusion genes introduced into chromaffin cells in primary culture. J Neurosci 1990; 10:520-30.
76. Seidl K, Unsicker K. The determination of the adrenal medullary cell fate during embryogenesis. Dev Biol 1989; 136:481-90.

77. Margolis FL, Roffi J, Jost A. Norepinephrine methylation in fetal rat adrenals. Science 1966; 154:275-6.
78. Wurtman RJ, Axelrod J. Control of enzymatic synthesis of adrenaline in the adrenal medulla by adrenal cortical steroids. J Biol Chem 1970; 241:2301-5.
79. Ciaranello RD, Jacobowitz D, Axelrod J. Effect of dexamethasone on phenylethanolamine N-methyltransferase in chromaffin tissue of the neonatal rat. J Neurochem 1973; 20:799-805.
80. Teitelman G, Joh TH, Park DH et al. Expression of the adrenergic phenotype in cultured fetal adrenal medullary cells: Role of extrinsic and intrinsic factors. Dev Biol 1982; 89:450-9.
81. Grothe C, Hofmann HD, Verhofstad AAJ et al. Nerve growth factor and dexamethasone specify catecholaminergic phenotype of cultured rat chromaffin cells: Dependence on developmental stage. Dev Brain Res 1985; 21:125-32.
82. Michelson AM, Anderson DJ. Changes in competence determine the timing of two sequential glucocorticoid effects on sympathoadrenal progenitors. Neuron 1992; 8:589-604.
83. Lillien LE, Claude P. Nerve growth factor is a mitogen for cultured chromaffin cells. Nature 1985; 317:632-4.
84. Naujoks KW, Korsching S, Rohrer H et al. Nerve growth factor-mediated induction of tyrosine hydroxylase and of neurite outgrowth in cultures of bovine adrenal chromaffin cells: dependence on developmental stage. Dev Biol 1982; 92:365-79.
85. Tischler AS, Riseberg JC, Hardenbrook MA et al. Nerve growth factor is a potent inducer of proliferation and neuronal differentiation for adult rat chromaffin cells in vitro. J Neurosci 1993; 13:1533-42.
86. Stemple DL, Mahanthappa NK, Anderson DJ. Basic FGF induces neuronal differentiation, cell division, and NGF dependence in chromaffin cells: A sequence of events in sympathetic development. Neuron 1988; 1:517-25.
87. Frodin M, Gammeltoft S. Insulin-like growth factors act synergistically with basic fibroblast growth factor and nerve growth factor to promote chromaffin cell proliferation. Proc Natl Acad Sci USA 1994; 91:1771-5.
88. Claude P, Parada IM, Gordon KA et al. Acidic fibroblast growth factor stimulates adrenal chromaffin cells to proliferate and to extend neurites, but is not a long-term survival factor. Neuron 1988; 1:783-90.
89. Anderson DJ. Molecular control of cell fate in the neural crest: the sympathoadrenal lineage. Ann Rev Neurosci 1993; 16:129-58.

CHAPTER 6

The Adrenal Medulla in MEN 2

Arthur S. Tischler and Ronald A. DeLellis

INTRODUCTION

The multiple endocrine neoplasia type 2 (MEN 2) syndromes are a fascinating group of proliferative disorders that classically involve both the adrenal medulla and thyroid C-cells. The 1961 report by Sipple,[1] then a surgery resident, of the more than coincidental association between pheochromocytoma and thyroid carcinoma focused attention on the entity now known as MEN 2A. Subsequent studies by Williams[2] established that the thyroid carcinomas were of the medullary type. Several papers shortly thereafter led to the recognition of MEN 2B and apparent variants of the two classic syndromes were subsequently described, as recently reviewed.[3] These variants include familial pheochromocytoma without medullary thyroid carcinoma, and familial medullary thyroid carcinoma without pheochromocytoma.

Pheochromocytomas occur in 30-70% of patients with MEN 2,[4] in contrast to their reported frequency of about 0.005% in unselected autopsies[5] and an annual incidence of about 0.8 per 100,000 person-years, excluding familial cases.[6] It is estimated that approximately 10% of adrenal pheochromocytomas occur in patients with familial syndromes,[5] most often MEN 2. In contrast to pheochromocytomas that are sporadic, those associated with MEN 2 are usually bilateral and/or multicentric.[7,8] However, the

Genetic Mechanisms in Multiple Endocrine Neoplasia Type 2, edited by Barry D. Nelkin.

two adrenals may be affected either synchronously or metachronously. In a prospective study of adrenal medullary tumors in 19 patients with MEN 2A by Gagel et al,[8] approximately 60% of patients presented with bilateral tumors. An additional 20% developed second tumors within four years after unilateral adrenalectomy. DeLellis et al[9] reported multicentric pheochromocytomas in 30% of patients undergoing unilateral adrenalectomy. Pheochromocytomas may occur prior to, concurrently with, or after medullary thyroid carcinomas. Their detection after C-cell abnormalities in most cases[8] might in part reflect the absence of a screening test as sensitive as the calcitonin assay.

Inconsistency of nomenclature is a potential source of confusion for those seeking to understand the pathobiology of MEN 2. Catecholamine-producing organs known as paraganglia are associated with both the sympathetic and parasympathetic nervous systems.[10] When multiple tumors occur in these organs within individuals or kindreds, they are usually confined to either the sympathetic or parasympathetic class. The adrenal medulla is the prototypical sympathetic paraganglion, while the prototypical parasympathetic paraganglion is the carotid body. With the exception of these two prototypes, most other paraganglia are microscopic. In the United States, most publications have followed the suggestion by Karsner[11] that the term "pheochromocytoma" be reserved for adrenal medullary tumors and that tumors of extra-adrenal paraganglia be called "paragangliomas." In other literature, however, especially in Europe and Japan, neoplasms of extra-adrenal sympathetic paraganglia are referred to as "extra-adrenal pheochromocytomas," particularly if they exhibit a positive chromaffin reaction or are associated with clinical evidence of catecholamine secretion. This inconsistency may obscure the fact that extra-adrenal paragangliomas are quite uncommon in most kindreds with MEN 2, although they may occur.[12,13] Typical MEN 2 patients thus contrast with the occasional pediatric or adult patients who may present with multiple sympathetic paragangliomas and have no relevant family history.[14,15] Up to 21 separate paragangliomas have been reported in such individuals.[14] Although approximately 25% of extra-adrenal sympathetic paragangliomas occur in combination with adrenal pheochromocytomas,[5] the differences between MEN 2 and other clinical syndromes suggest that physiological signals driving proliferation of adrenal chromaffin cells are to some extent different from those driving proliferation of other sympathetic paraganglia.

Identification of abnormalities in the ret proto-oncogene as the molecular common denominator of the MEN 2 syndromes,[16-18] has provided a foundation for investigating the cellular mechanisms involved in neoplastic transformation. Ret has been shown to be consistently expressed by pheochromocytomas and medullary thyroid carcinomas.[19-21] Ret transcripts are also found in normal thyroid and adrenal glands.[20-22] Pheochromocytomas from patients with MEN 2A can be retrospectively analyzed for ret mutations using DNA retrieved from paraffin sections.[23] One pressing question that now needs to be addressed is whether ret mutation alone is sufficient to cause adrenal medullary neoplasia. Circumstantial evidence that this is the case is provided by the findings of Lindor et al[24] of ret mutations in a subset of sporadic pheochromocytomas. Moreover, Santoro et al[25] have shown that transfection of NIH 3T3 fibroblasts with mutated ret results in a transformed phenotype. However, it must be borne in mind that while all adrenal cells in MEN 2 patients carry mutated ret, only a small percentage transform, and that in most sporadic pheochromocytomas ret thus far appears to be normal. The transformation of NIH 3T3 cells shows that mutated ret may function as a dominant oncogene under appropriate circumstances. However, NIH 3T3 cells are immortal and might therefore have already sustained genetic damage, such as deletion of a tumor suppressor gene, that would ordinarily initiate transformation in vivo. A second important question is to what extent do epigenetic influences contribute to the variable expression of the MEN 2 phenotype. If their contribution is substantial, the possibility of novel therapeutic or prophylactic interventions is raised. Considered together, clinical data, morphological analyses of affected adrenal glands and insights from animals models suggest that the development of MEN 2 involves a complex interplay of genetic and epigenetic factors.

THE ADRENAL MEDULLA IN MEN 2

Proliferative changes in the adrenal medulla of patients with MEN 2 comprise a morphologic spectrum from diffuse hyperplasia to pheochromocytoma.[9, 26] In between is an entity known as nodular hyperplasia, consisting of small nodules on a diffusely hyperplastic background. These changes are generally considered to represent a progression, starting with diffuse hyperplasia.[9] However, the amount of diffuse hyperplasia in adrenals with nodules is highly variable, and in some cases multiple nodules may exist without apparent diffuse hyperplasia (Fig. 6.1).

Discrimination of diffuse adrenal medullary hyperplasia from normal anatomic variations in the amount of medullary tissue may be difficult, particularly when the hyperplasia is mild. Medullary tissue is usually distributed exclusively in the central portion of the head and body of the normal adrenal gland.[27] Suspicion of hyperplasia is raised when the medulla extends into the tail and/or into both alae in the body.[27] However, over many years of dissecting adrenals for cell culture, one of us (AST) has occasionally observed adrenal glands from individuals without apparent endocrine abnormalities that contain medulla in the alae. Some of these cases may have represented undiagnosed examples of the apparently sporadic adrenal medullary hyperplasia described by Rudy et al.[28] Hyperplasia may also be suspected when the area of medulla to cortex greatly exceeds 1:4 in any individual section, because the average ratio in the body of normal adrenals is 1:5.[9] A definite diagnosis, however, can be established only by morphometric evaluation. The most reliable method is calculation of total adrenal medullary weight and volume from representative sections of multiple consecutive blocks of the entire gland.[9] This approach avoids potential error caused by atrophy or hyperplasia of the cortex, or by anatomic variation in the shape of the medulla. Medullary weight calculated for the normal adult human adrenal medulla is between 0.3-0.5g,[9,28] about 10% of the weight of the cortex. In adrenal glands from patients with MEN 2, the diffusely expanded medullas may weigh in excess of 1.6g.[9]

A number of cytological changes often accompanies the development of diffuse and nodular adrenal medullary hyperplasia. These include cellular pleomorphism and hypertrophy, prominent mitoses, an open chromatin pattern, enlarged nucleoli and the presence of multinucleated cells and giant cells with bizarre, hyperchromatic nuclei (Figs. 6.2 and 6.3).[9,26,28] The changes are consistent with both functional and mitogenic stimulation accompanied by accrual of genetic damage. It must, however, be remembered that chromaffin cells from normal individuals may exhibit some pleomorphism and that cell size, nuclear morphology and chromatin pattern may be affected by tissue fixation and processing.

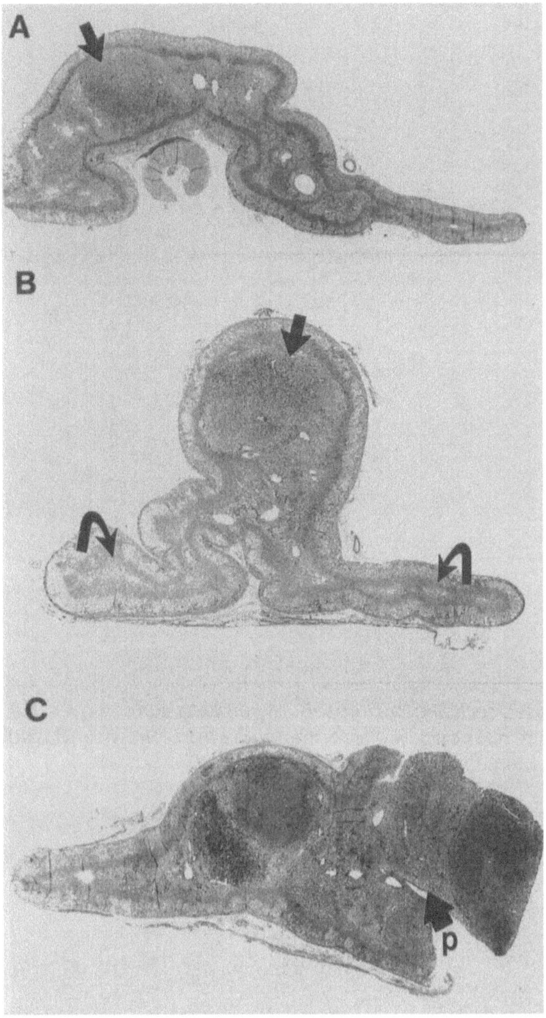

Fig. 6.1. Low magnification of cross sections of hematoxylin- and eosin-stained adrenal glands from patients with MEN 2, who underwent adrenalectomy after detection of abnormalities in levels of circulating catecholamines.[9] In panel (A.)The medulla in panel A is diffusely widened, with one large nodule (arrow) and scattered ill-defined areas of increased cellularity. Panel (B) shows one prominent medullary nodule (straight arrow) and very little evidence of diffuse medullary hyperplasia, except for small amounts of medullary tissue in the alae (curved arrows). Panel (C) shows both diffusely widened medullary tissue in the alar ridge at left and multiple nodules of varying sizes, with a large pheochromocytoma (p) at right. Panels (A) and (B) reproduced with permission from DeLellis et al. Am J Pathol 1976; 83:177-196.[9]

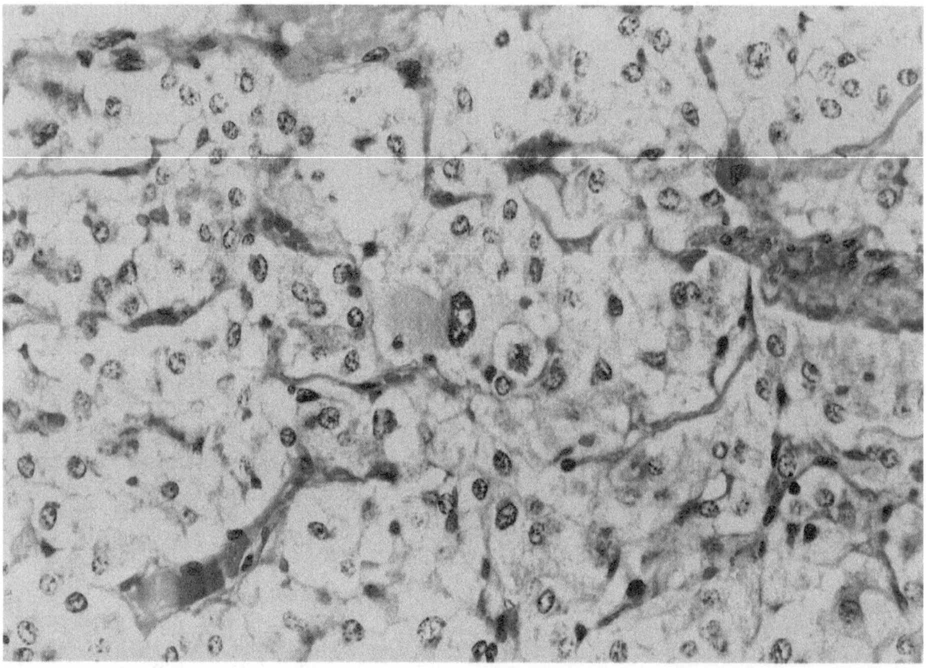

Fig. 6.2. Higher magnification of H&E-stained section of a diffusely hyperplastic adrenal medulla showing mild cellular pleomorphism and one bizarre giant cell (center) with a markedly hyperplastic nucleus. Such giant cells are rarely encountered in control adrenals.[26] (x 400)

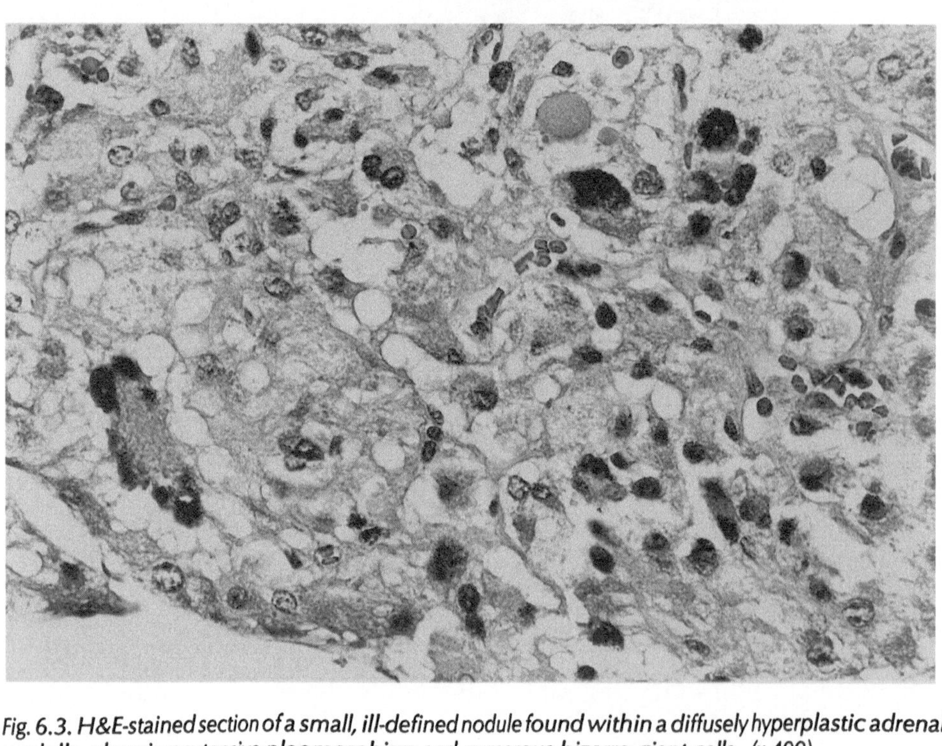

Fig. 6.3. H&E-stained section of a small, ill-defined nodule found within a diffusely hyperplastic adrenal medulla, showing extensive pleomorphism and numerous bizarre giant cells. (x 400)

We have recently employed the monoclonal antibody, MIB 1, to evaluate cell proliferation in non-nodular areas of adrenals from MEN 2 patients. MIB-1 is directed against the Ki-67 proliferation antigen, but differs from other Ki-67 antibodies in that it may be used in paraffin sections (Fig. 6.4).[29] Preliminary data suggest that substantial increases in chromaffin cell proliferation may precede or accompany the development of pheochromocytomas. However, the greatest proliferation appears to be found in sections in which the medulla visually appears widened. Reasons for this intraglandular variation in chromaffin cell proliferation remain unknown. There also appears to be substantial overlap between control adrenals and those from patients with MEN 2. The former usually contain less than one MIB-1-labeled chromaffin cell nucleus per square millimeter of medulla, while MEN 2 adrenals may contain eight or more. However, MEN 2 adrenals may fall within the normal range.

Adrenal medullary hyperplasia is accompanied by an increased ratio of epinephrine to norepinephrine.[9,26] Other functional changes

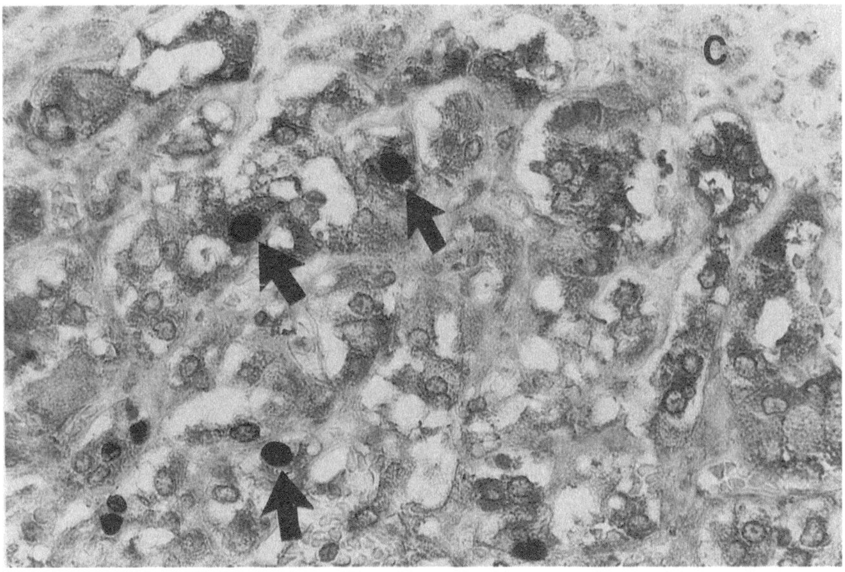

Fig. 6.4. Diffusely hyperplastic adrenal gland double-stained for tyrosine hydroxylase (TH) and for the nuclear proliferative antigen Ki-67 using monoclonal antibody MIB-1. Dark cytoplasmic staining for TH, the rate-limiting enzyme in catecholamine synthesis, discriminates chromaffin cells from other cell types,[34] including a small amount of cortex (c) at top. Multiple nuclei reactive with MIB-1 (arrows) reveal an unusually active focus of chromaffin cell proliferative activity. (x 400)

have not been studied extensively. In an interesting recent case, we have observed immunoreactive calcitonin (CT) in scattered chromaffin cells in an adrenal that also contained a pheochromocytoma (Fig. 6.5). Although CT in the adrenal in MEN 2 patients has been reported to be indicative of metastatic medullary thyroid carcinoma (MTC),[30] our findings and those reported by others[31] suggest that CT, which is not normally produced in the adrenal, may, at least occasionally, be ectopically produced by chromaffin cells.

The dividing line between nodular hyperplasia and pheochromocytoma is unknown. Lack of encapsulation or of reticulin compression in the adjacent adrenal have been suggested as arbitrary criteria for hyperplasia,[28] but large, encapsulated tumors must logically begin small and uncapsulated. Baylin et al[32] have shown that at least some of the small nodules are monoclonal, and monoclonality might be interpreted as evidence that any nodule, regardless of size, has crossed the line into neoplasia. However, chromaffin cells are heterogeneous in many respects. The possibility that individual cells are able to proliferate in response to certain

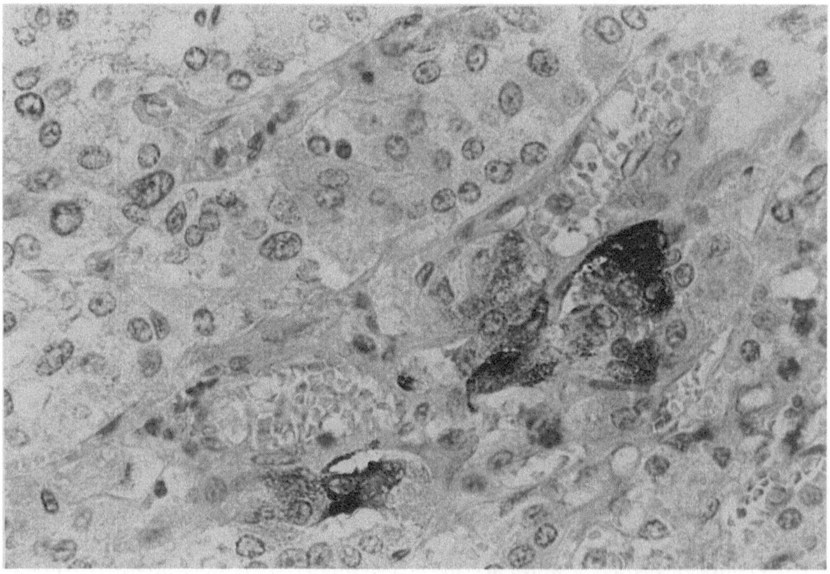

Fig. 6.5. Scattered adrenal medullary cells immunoreactive for calcitonin (with rabbit antiserum from DAKO Corporation, used at a dilution of 1:3000) in minimally hyperplastic gland. A pheochromocytoma was present in another field. (x 400)

signals and thus generate true hyperplastic nodules that are monoclonal has not been ruled out.

Aside from their multicentricity, pheochromocytomas that occur in association with MEN 2 are similar to their sporadic counterparts. They metastasize in fewer than 5% of patients and there are no unequivocal criteria for determining malignancy other than the presence of metastases. The tumors consistently appear to produce epinephrine. Gagel et al[8] have reviewed clinical data from 18 patients with MEN 2 who underwent adrenalectomy (ten bilateral and eight unilateral adrenalectomies) over an 18-year period and found that 24-hour urinary epinephrine levels or ratios of epinephrine to norepinephrine were abnormal in all but one patient with a diagnosis of pheochromocytoma. Immunohistochemical studies of catecholamine-synthesizing ability of MEN 2 adrenals have recently been conducted in our laboratory using antibodies against catecholamine biosynthetic enzymes. This methodology represents a radical advance for histopathological studies of catecholamine-producing tissues because it enables the synthesis of specific catecholamines to be inferred from paraffin sections,[33, 34] thereby eliminating the need for analyses that require fresh tissue. In the normal human adrenal, almost all chromaffin cells show strong immunoreactivity for phenylethanolamine-N-methyltransferase (PNMT), the enzyme that synthesizes epinephrine from norepinephrine.[10] In MEN 2 adrenals, normal-appearing and diffusely hyperplastic medulla show comparable staining for PNMT. In contrast, we have found that nodules and pheochromocytomas, while PNMT-positive, may show less immunoreactivity than adjacent normal or diffusely hyperplastic medulla (Fig. 6.6). In light of these observations, it is important to realize that the ratio of circulating E/NE is not a direct reflection of the E/NE ratio in the adrenal. It does reflect the amount of adrenal medullary tissue relative to other catecholamine-synthesizing tissues, which produce little or no circulating epinephrine.

ANALOGIES TO MEN 2 IN RODENT MODELS

Several rodent models share features with MEN 2 syndromes and might provide insights into the pathobiology of these disorders.

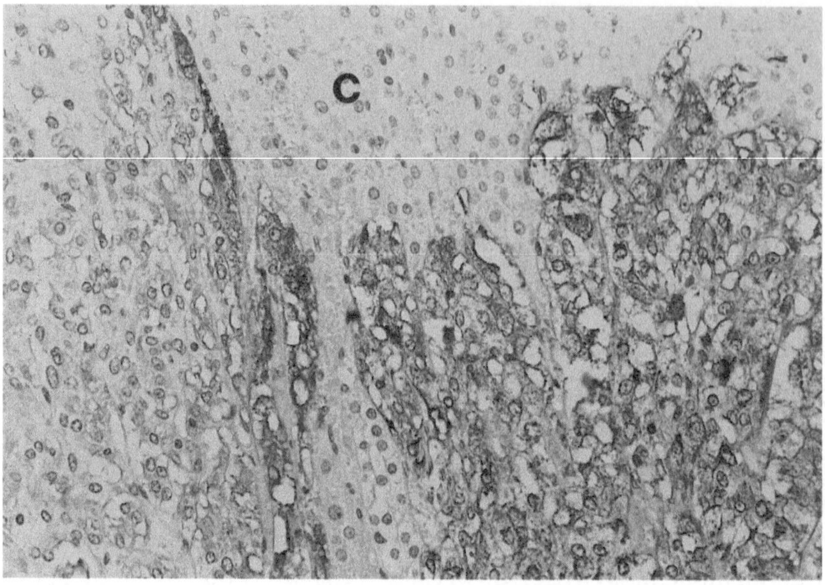

Fig. 6.6. Immunohistochemical staining for PNMT in an adrenal with a small pheochromocytoma at left and diffusely hyperplastic medulla at right. A small amount of intervening cortex (c) is present at top and center. Immunoreactivity for PNMT is less intense in the pheochromocytoma than in the diffuse hyperplasia. (x 200)

The Rat Pheochromocytoma Model and Chromaffin Cell Proliferation

In contrast to their rarity in humans without MEN 2, adrenal medullary hyperplasia and pheochromocytomas arise frequently in the rat, either spontaneously in the course of aging or in response to a wide variety of drugs, dietary factors, toxins and other agents. The incidence of these lesions in rats is variable and is dependent on both genetic and epigenetic factors, as recently reviewed.[35-37] The former include the animals' strain and sex. Pheochromocytomas occur in up to 80% of Wistar and Wistar-derived rats in lifetime studies, most frequently in males. Epigenetic factors exacerbate the proclivity of any given strain to develop these tumors. Pheochromocytomas occur most frequently in older animals and their incidence is increased by a long list of agents including reserpine, nicotine, growth hormone, estrogen, nonsteroid anti-inflammatory agents, phosphodiesterase inhibitors, anti-thyroid drugs, retinoids and neuroleptics. Also implicated are sugars and sugar alcohols that affect absorption of dietary calcium, miscellaneous drugs and toxins and radiation.[35-37] The diversity of these agents, together with the lack of evidence, in most instances, of the ability to cause DNA damage suggests that many of the agents influence the adrenal medulla indirectly rather than by a direct carcinogenic effect.

In vivo investigations employing mitotic counts and BrdU labeling to study possible mechanisms of adrenal medullary carcinogenesis has shown that proliferation of rat chromaffin cells is increased by administration of reserpine[37,38] and decreased by unilateral adrenal denervation.[38] Reserpine is known to deplete catecholamine stores and to reflexly increase the activity of the splanchnic nerve endings innervating the adrenal medulla to stimulate both secretion and synthesis of catecholamines and other secretory granule constituents.[39] The effects of reserpine and denervation on chromaffin cell proliferation in the rat therefore suggest that neurally derived signals might also normally regulate chromaffin cell number to meet changing physiological needs. This regulation might in turn provide a common mechanism permitting pharmacologically diverse substances to produce adrenal medullary tumors by increasing cell turnover and providing a proliferative backdrop on which genetic damage might occur. We believe that this damage might be mediated by oxidative mechanisms involving interactions of catecholamines with molecular oxygen.[40,41]

Cell culture studies have been designed to mimic the effects of innervation by direct pharmacological activation of intracellular signaling pathways normally activated by neurotransmitters in adrenal medullary nerve endings. In cell cultures prepared from adult rats, chromaffin cell proliferation is stimulated either by the peptide growth factors NGF[42] and FGF,[43] or by activators of adenylate cyclase or protein kinase C.[42] Differing susceptibilities to inhibitors and potentiators suggest that growth factors, cyclic AMP-dependent protein kinases and protein kinase C act via partially distinct and partially overlapping signaling pathways. Moreover, activators of adenylate cyclase inhibit mitogenic responses to NGF or FGF.[42]

Considered together, current in vivo and in vitro data from the rat model suggest that during normal development, neurally derived signals supersede growth factors in regulating proliferation of rat chromaffin cells by selectively inhibiting or co-opting portions of growth factor signaling pathways. This switch from humoral to neural regulation might provide a means of fine-tuning proliferation in adult life through both positive and negative control.[42] Rat pheochromocytomas might result either from constitutive activation or defective inhibition of mitogenic signaling pathways, and interspecies differences in mitogenic signaling might contribute to the different frequencies of adrenal medullary tumors. This hypothesis is supported by the fact that human chromaffin cells do not proliferate in response to agents that are mitogenic for chromaffin cells from rats.[44] The rat model suggests that MEN 2 tyrosine phosphorylation by mutated ret might provide a proliferative backdrop that facilitates other genetic damage required for neoplastic transformation.

Mouse Models of Oncogene Activation and Anti-Oncogene Deletion

In contrast to rats, mice normally exhibit a very low frequency of pheochromocytomas.[45] Recently developed transgenic and knockout mouse models demonstrate that there is more than one way to produce a pheochromocytoma.

Three out of four strains of transgenic mice overexpressing the c-mos proto-oncogene have been shown to frequently develop pheochromocytomas and/or medullary thyroid carcinomas.[46] The c-mos gene product is normally involved in regulation of mitosis and meiosis. Its oncogenic role has been postulated to involve aberrant cell cycle regulation. Pheochromocytomas develop in no more than about 60% of the transgenic animals, and occur after a long latency, suggesting that multiple steps are required for neoplastic transformation.[46]

Pheochromocytomas have been shown to occur frequently in mice with knockout mutations of the neurofibromatosis gene (Nf1)[47] or the retinoblastoma gene (Rb).[48] The Rb knockout is also associated with medullary thyroid carcinomas.[48] The protein encoded by Nfl contains GTPase-activating domains that control the activity of ras proteins in intracellular signaling, while Rb is a multifunctional tumor suppressor gene. It is of interest not only that knockouts of these unrelated tumor suppressor genes can independently lead to the development of pheochromocytomas but that they can also affect the phenotype of the tumors. Pheochromocytomas associated with the Nfl knockout are usually large and often resemble human pheochromocytomas in expressing immunoreactive PNMT.[49] In contrast, the pheochromocytomas associated with the Rb knockout are usually small and PNMT-negative (Tischler and Williams, unpublished data).

The importance of the transgenic and knockout mouse models may not be so much to focus on specific genes potentially involved in MEN 2 as to emphasize potential genetic interplay and interchangeability. The persistent association of adrenal medullary and C-cell tumors in animals with different genetic abnormalities is also of interest in suggesting that there is considerable overlap in mitogenic signaling mechanisms in these two tissues of neural crest origin.

The expression of mos has been identified in only one of nine tumors examined from MEN 2 patients.[46] Abnormalities of *NF1* expression have been demonstrated in pheochromocytomas associated with neurofibromatosis.[50] Pheochromocytomas from MEN 2 patients are currently being studied in our laboratory, and *NF1* abnormalities have thus far not been detected.

An additional potentially important use of the mouse models is to clarify the role of genetic background in expression of MEN 2 phenotypes. In both the mos and Nf1 models, pheochromocytomas occur only when animals bearing the transgene or knockout mutation are outbred to create F1 hybrids of mixed genetic background.[46,49]

SUMMARY AND PERSPECTIVES

The adrenal medullary changes in MEN 2 support a hypothetical model in which increased cell proliferation driven directly or indirectly by mutated ret facilitates the accrual of additional genetic insults, ultimately leading to pheochromocytomas. The amount of diffuse hyperplasia that coexists with pheochromocytomas in MEN 2 adrenals is highly variable, perhaps reflecting how long it takes for a critical gene to be appropriately damaged. Functional alterations accompany morphological changes in diffuse hyperplasia, suggesting that activated ret may affect multiple intracellular signaling pathways.

A situation apparently comparable to MEN 2 exists in laboratory rats, which have a much higher rate of chromaffin cell proliferation and a much higher frequency of pheochromocytomas than normal humans. Although the amount of cell proliferation in MEN 2 adrenals is less than in normal rat adrenals, allowance must be made for the accrual of genetic damage over the life expectancy of 70 or so years of humans, versus the three years of rats. Data from transgenic and knockout mouse models have shown that alterations in several unrelated genes can lead to pheochromocytomas. More than one gene might therefore also be able to produce tumors in MEN 2. Ret itself has thus far not been shown to be implicated in any of the rodent models.

Further understanding of the adrenal medullary abnormalities in MEN 2 will require knowledge of the mechanisms that regulate proliferation of normal and neoplastic human chromaffin cells and of the involvement of ret in these mechanisms. If mutated ret is found to be involved in mitogenic signaling, it will be important to determine whether this is also true of its normal counterpart. Although proliferation of rat chromaffin cells appears at least in part to be neurally mediated, it is not known whether this is also the case in humans, or whether slide-to-slide variation in chromaffin cell proliferation in MEN 2 adrenals is due to lack of uniform innervation. It is also unknown whether exogenous factors such as drugs or diet can affect chromaffin cell proliferation in humans, as they can in rats. Neither normal nor neoplastic human chromaffin cells proliferate in cell cultures in response to substances that are mitogenic for chromaffin cells from rats, suggesting that different types of regulation also exist in vivo.

Acknowledgment

Research discussed in this chapter was supported by NIH grants CA27808 and CA48017 and a grant from the ILSI Risk Science Institute.

References

1. Sipple JH. The association of pheochromocytoma with carcinoma of the thyroid gland. Am J Med 1961; 31:163-6.
2. Williams ED. Histogenesis of medullary carcinoma of the thyroid. J Clin Pathol 1966; 19:114-8.
3. DeLellis RA. Multiple endocrine neoplasia syndromes revisited. Clinical, morphologic and molecular features. Lab Invest 72:1-12.
4. Page DL, DeLellis RA, Hough AJ Jr. Tumors of the adrenal. In: Atlas of Tumor Pathology, 2nd ed. Fasc 23, Armed Forces Institute of Pathology, Washington, DC:1986.
5. Manger WM, Gifford RW. Pheochromocytoma. New York: Springer-Verlag 1977:44-8.

6. Beard CM, Sheps SG, Kurland LT et al. Occurrence of pheochromocytoma in Rochester, Minnesota, 1950-1979. Mayo Clinic Proceedings 1983; 802-4.
7. Evans DB, Lee JE, Merrell RC et al. Adrenal medullary disease in multiple endocrine neoplasia type 2. Appropriate management. Endocrinol Metabol Clin North Am 1994; 23:167-76.
8. Gagel RF, Tashjian AH Jr, Cummings T et al. Impact of prospective screening for multiple endocrine neoplasia type 2. 1988; N Engl J Med 318:478-84.
9. DeLellis RA, Wolfe HJ, Gagel RF et al. Adrenal medullary hyperplasia. Am J Pathol 1976; 83:177-96.
10. Tischler AS. Paraganglia. In: Sternberg S, ed. Histology for Pathologists. New York: Raven Press, 1992:363-97.
11. Karsner HT. Tumors of the Adrenal. In: Atlas of Tumor Pathology, Vol Fasc 29. Armed Forces Institute of Pathology, Washington, DC, 1950.
12. Lips CMJ, Minder WH, Leo JR et al. Evidence of multicentric origin of the multiple endocrine neoplasia syndrome type 2A (Sipple's syndrome) in a large family in the Netherlands. Am J Med 1978; 64:569-78.
13. Marks AD, Channick BJ. Extra-adrenal pheochromocytoma and medullary thyroid carcinoma with pheochromocytoma. Arch Int Med 1974; 134:1106-9.
14. Kawai K, Kimura S, Miyamoto J et al. A case of multiple extra-adrenal pheochromocytomas. Endocrinol Jpn 1979; 26:693-6.
15. Karasov RS, Sheps SG, Carney AJ. Paragangliomatosis with numerous catecholamine-producing tumors. Mayo Clinic Proceedings 1982; 57:590-5.
16. Gardner E, Papi L, Easton DF et al. Genetic linkage studies map the multiple endocrine neoplasia type 2 loci to a small interval on chromosome 10q11.2. Hum Mol Genetics 1993; 2:241-6.
17. Lairmore TC, Dou S, Howe JR et al. A 1.5 megabase yeast artificial chromosome contig from human chromosome 10q11.2 connecting three genetic loci (RET, D10S94 and D10S102) closely linked to the MEN 2A locus. Proc Natl Acad Sci USA 1993; 90:492-6.
18. Mole SE, Mulligan LM, Healey CS et al. Localization of the gene for multiple endocrine neoplasia type 2A to a 480kb region in chromosome band 10q11.2. Hum Mol Genetics 1993; 2:247-52.
19. Nagao M, Ishizaka Y, Nakagawara A et al. Expression of ret protooncogene in human neuroblastomas. Jap J Cancer Res 1990; 81:309-12.
20. Santoro M, Rosati R, Grieco M et al. The ret proto-oncogene is consistently expressed in human pheochromocytomas and thyroid medullary carcinomas. Oncogene 1990; 5:1595-8.
21. Miya A, Yamamoto M, Morimoto H et al. Expression of the ret

proto-oncogene in human medullary thyroid carcinomas and pheochromocytomas of MEN 2A. Henry Ford Hosp Med J 1992; 40:215-19.
22. Pachnis V, Mankoo B, Costantini F. Expression of the c-ret proto-oncogene during mouse embryogenesis. Development 1993; 119:1005-17.
23. Komminoth P, Kunz E, Hiört O et al. Detection of ret proto-oncogene point mutations in paraffin-embedded pheochromocytoma specimens by non-radioactive single-strand confirmation polymorphism analysis and direct sequencing. Am J Pathol 1994; 144:922-29.
24. Lindor NM, Honchel R, Khsla S et al. Mutations in the ret proto-oncogene in sporadic pheochromocytomas. J Clin Endocrin Metab 1995; 80:627-29.
25. Santoro M, Carlomagno F, Romano A et al. Activation of ret as a dominant transforming gene by germline mutations of MEN 2A and MEN 2B. Science 1995; 267:381-383.
26. Carney JA, Sizemore GW, Tyce GM. Bilateral adrenal medullary hyperplasia in multiple endocrine neoplasia, type 2. Mayo Clinic Proceedings 1975; 50:3.
27. Neville AM. The adrenal medulla. In: Symington T, ed. Functional Pathology of the Human Adrenal Gland. 1969.
28. Rudy FR, Bates RD, Cimorelli AJ et al. Adrenal medullary hyperplasia: A clinicopathologic study of four cases. Human Pathol 1980; 26:131-34.
29. DeLellis RA. Does the evaluation of proliferative activity predict malignancy or prognosis in endocrine tumors? Human Pathol 1995; 26:131-34.
30. Mendelsohn G, Baylin SB, Eggleston JC. Relationship of metastatic medullary thyroid carcinoma to carcinoma content of pheochromocytomas: An immunohistochemical study. Cancer 1980; 45:498-502.
31. O'Connor DT, Frigon RP, Deftos W. Immunoreactive calcitonin in catecholamine storage vesicles of human pheochromocytoma. J Clin Endocrin Metab 1983; 56:582-85.
32. Baylin SB, Gann DS, Shu SH. Clonal origin of inherited medullary thyroid carcinoma and pheochromocytoma. Science 1976; 193:321-23.
33. Lloyd RV, Sisson K, Shapiro B. Histochemical localization of epinephrine, norepinephrine, catecholamine-synthesizing enzymes and chromogranin in neuroendocrine cells and tumors. Am J Pathol 1986; 125;45-54.
34. Tischler AS. Triple immunohistochemical staining for bromodeoxyuridine and catecholamine biosynthetic enzymes using microwave antigen retrieval. J Histochem Cytochem 1995; 43:1-4.
35. Tischler AS, DeLellis RA. The rat adrenal medulla. II. Proliferative

lesions. J Amer Coll Toxicol 1988; 7:23-44.
36. Tischler AS, Coupland RE. Age-related changes in the structure and function of the rat adrenal medulla. In: Mohr U, Dungworth DL, Capen CL, eds. Pathobiology of the Aging Rat. Vol. 2. Washington DC. ILSI Press, 1994:245-68.
37. Tischler AS, Ruzicka LA, Donahue SR et al. Chromaffin cell proliferation in the adult adrenal medulla. Int J Devel Neurosci 1989; 7:439-448.
38. Tischler AS, McClain RM, Childers H et al. Neurogenic signals regulate chromaffin cell proliferation and mediate the mitogenic effect of reserpine in the adult rat adrenal medulla. Lab Invest 1991; 65:374-76.
39. Sietzen M, Schober M, Fischer-Colbrie R et al. Rat adrenal medulla. Levels of chromogranins, enkephalins, dopamine beta-hydroxylase and the amine transporter are changed by nervous activity and hypophysectomy. Neuroscience 1987; 22:131-39.
40. Graham DG. On the origin and significance of neuromalanin. Arch Pathol Lab Med 1979; 103:359-62.
41. Baez S, Segura-Aguilar J. Formation of reactive oxygen species during one-electron reduction of noradrenochrome catalyzed by NADPH-cytochrome P-450 reductase. Redox Report 1994; 1:65-70.
42. Tischler AS, Riseberg JC, Cherington V. Multiple mitogenic signaling pathways in chromaffin cells: A model for cell cycle regulation in the nervous system. Neurosci Lett 1994; 168:181-84.
43. Mahanthappa NK, Gage FG, Patterson PH. Adrenal chromaffin cells as multipotential neurons for autografts. Prog Brain Res 1990; 82:33-39.
44. Tischler AS, Riseberg JC. Different responses to mitogenic signals by adult rat and human chromaffin cells *in vitro*. Endocrine Pathol 1993; 4:15-19.
45. Tischler AS, Sheldon W. The adrenal medulla in the aging mouse. In: ILSI Monographs on Pathobiology of Aging Animals. The Aging Mouse. Washington DC: ILSI Press, in press.
46. Schulz N, Propst F, Rosenberg MP et al. Pheochromocytomas and C-cell thyroid neoplasms in transgenic *c-mos* mice: A model for the human multiple endocrine neoplasia type 2 syndrome. Cancer Res 1992; 52:450-55.
47. Jacks T, Shih TS, Schmitt EM et al. Tumorigenic and developmental consequences of a targeted Nf1 mutation in the mouse. Nature Genetics 1994; 7:353-61.
48. Williams BO, Schmitt EM, Remington L et al. Extensive contribution of Rb-deficient cells to adult chimeric mice with limited histopathological consequences. EMBO Journal 1994; 13:4251-59.
49. Tischler AS, Shih TS, Williams BO et al. Characterization of pheochromocytomas in a mouse strain with a targeted disruptive mutation of neurofibromatosis gene Nf1. Endocrine Pathol 1996; 6:323-35.

50. Gutmann DH, Cole JL, Stone WJ et al. Loss of neurofibromin in adrenal gland tumors from patients with neurofibromatosis type 1. Genes, Chromosomes, Cancer 1994; 10:55-58.

CHAPTER 7

Neuronal Properties of Thyroid C-Cell Tumor Lines

Andrew F. Russo and Thomas M. Lanigan

ABSTRACT

Thyroid C-cells present a paradoxical phenotype of neuronal properties in multiple endocrine neoplasia type 2 (MEN 2). The neuronal phenotype is in contrast to the primarily endocrine nature of the parental C-cells. What can account for this transdifferentiation? In this chapter, we will try to answer that question by comparing the neuronal features of C-cell lines with those of normal C-cells treated with nerve growth factor. From that perspective, we will then present a model proposing that ret tyrosine kinase activity is responsible for the neuronal properties in C-cell tumors. Interestingly, the neuronal phenotype is partly reversible. Overexpression of the ras oncogene or treatment with glucocorticoids inhibits both cell growth and induces a more endocrine state. An attractive hypothesis is that these agents might be decreasing ret activity, although this remains to be tested. The utility of thyroid C-cell lines for identification of differentiation and growth genes will be exemplified by description of a novel splice product of the $G_s\alpha$ transcript that is selectively expressed in a subset of neuroendocrine and neuronal cells. Finally, the applicability of C-cell lines for studying fundamental transcriptional control

Genetic Mechanisms in Multiple Endocrine Neoplasia Type 2, edited by Barry D. Nelkin.

mechanisms will be described using the calcitonin/CGRP gene as a model. In summary, the induction and repression of neuronal properties in C-cell tumors provide a useful system for studying the genetic mechanisms underlying gene expression and differentiation in MEN 2.

INTRODUCTION

One of the manifestations of multiple endocrine neoplasia type 2 (MEN 2) is medullary thyroid carcinoma (MTC). These tumors arise from thyroid C-cells (parafollicular cells). C-cells are generally endocrine in nature and are distinguished by their expression of the hormone, calcitonin.[1,2] As such, their physiological function is primarily to sense the levels of extracellular calcium in order to help maintain calcium homeostasis.[3,4] However, they do have some intrinsic neuronal properties, such as serotonin biosynthesis and a neuronal serotonin binding protein.[5,6] Interestingly, C-cells show phenotypic plasticity with the ability to acquire even more neuronal features when placed in primary cultures or upon tumorigenesis, as discussed in this chapter.

The neuronal plasticity of C-cells can be rationalized by their origin from the neural crest. The neural crest is a transitory structure during embryogenesis that gives rise to a wide variety of cell types.[7] A hallmark of the neural crest is the flexibility of the progenitor cells to follow differentiation pathways determined to a large extent by the local microenvironment. C-cells arise from the vagal subdivision of the cranial crest, which also yields enteric neurons. This shared ontogeny, along with the neuronal plasticity and marker studies, have led to the hypothesis that C-cells and enteric neurons arise from a shared progenitor,[8-11] analogous to endocrine chromaffin cells and sympathetic neurons (Fig. 7.1).[12] The neuronal features of the C-cell lines may reflect the expression of genes that are otherwise repressed when the cells are within the normal context of the thyroid. Supporting this possibility, Jacobs-Cohen et al[13] have shown that triiodothyronine partly represses the development of a neuronal phenotype in primary C-cells. This paradigm is also analogous to the effect of the adrenal environment on chromaffin cells, which, interestingly, are also neoplastic targets in MEN 2, as discussed in Chapter 6.

We and others have used C-cells and C-cell lines as model systems for studying the mechanisms of neural crest differentia-

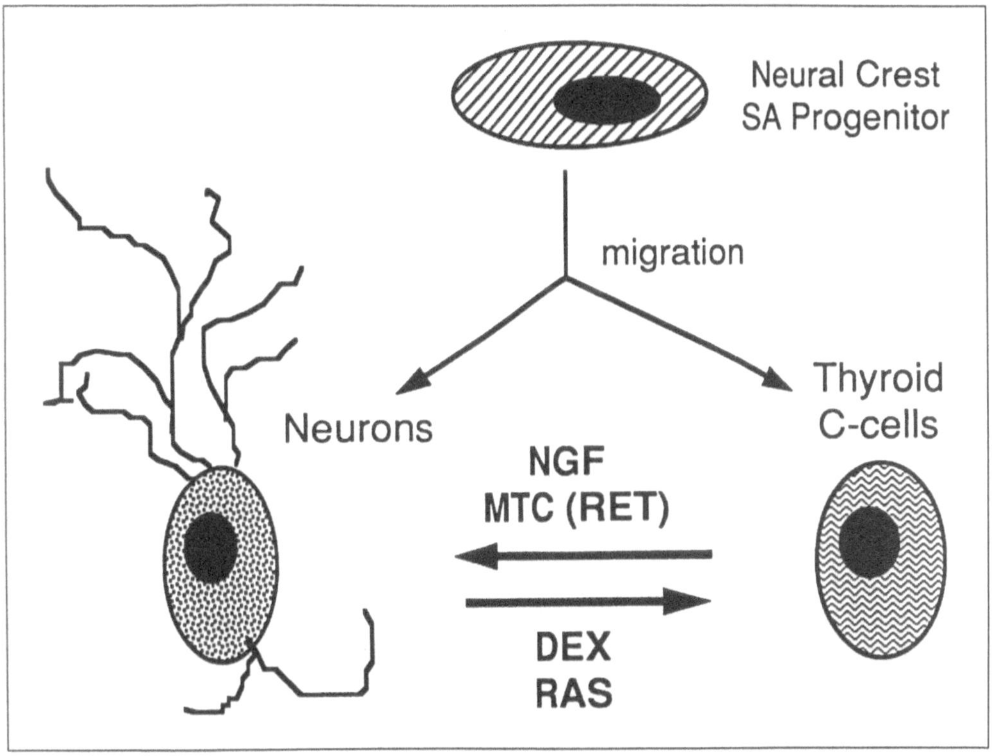

Fig. 7.1. Reversible neuronal transdifferentiation of thyroid C-cells. The sympathoadrenal progenitor in the vagal neural crest has been proposed to give rise to either thyroid C-cells or neurons, depending on the embryonic migration pathway.[11] Transdifferentiation of adult C-cells to a neuronal phenotype can be induced by nerve growth factor (NGF) treatment of primary C-cells and upon tumorigenesis in medullary thyroid carcinomas (MTC) in which the ret proto-oncogene has been activated. The neuronal differentiation of the MTC cell lines can be repressed to favor a more endocrine C-cell state by the glucocorticoid dexamethasone (DEX) or expression of the Harvey ras oncogene.

tion and gene regulation. Cell lines provide certain advantages towards understanding the molecular mechanisms that underlie complex regulatory events.[14] In this chapter, we will focus on the neuronal features of two C-cell lines, CA77[15] and TT,[16] that were isolated from rat and human MTCs, respectively. We will speculate on how ret tyrosine kinase activity may be contributing to their phenotype. As discussed in several chapters in this book, ret mutations have been genetically linked to MEN 2A and 2B (see Chapter 2),[17-20] as well as Hirschsprung's disease,[21,22] and ret is a key player in the terminal differentiation step of enteric neurons.[23] It should be noted that the c-mos proto-oncogene[24] and SV40 T-antigen[25] can also cause C-cell hyperplasia in transgenic mice, although only ret mutations have been associated with MEN-2. The reversible differentiation of CA77 and TT cells to an endocrine phenotype will then be described in the context of how this may provide insight into ret regulation. Finally, the application of these cell lines for differential cDNA hybridization and transcription studies to identify genes and regulatory factors in MEN-2 will be discussed.

NEURONAL PROPERTIES AND DIFFERENTIATION IN C-CELL CULTURES

The C-cell lines differ in the degree of their neuronal phenotypes.[26] For example, under normal culture conditions, the CA77 cells have a neuritic morphology and neurofilaments,[10] while the TT cell line is fusiform with short fibroblast-like processes.[16] Yet under the proper conditions, even the TT cells become more neuronal, and hence both CA77 and TT C-cell lines provide useful model systems. Furthermore, the observation that two out of three MTC tumors express neurofilament immunoreactivity[27] and 60% express neuron-specific enolase[28] supports the conclusion that neuronal differentiation is a general feature of C-cells upon tumorigenesis.

The CA77 C-cell line was generated by Bernard Roos in the early 1980s from a serially passaged rat MTC.[15] It has extensive neuronal properties (Table 7.1). The morphology is immediately visible as different from the round parental C-cells, with neurites well over 100 μm long on CA77 cells (Fig. 7.2).[10] When grown on a laminin substratum, about 70% of cells have neurites greater than 2 cell lengths after one day of growth. Growth cones and both clear and dense core vesicles can be detected by electron microscopy in both the cell bodies and processes. The cells have been shown to express all three neurofilaments, NF-L, NF-M and NF-H. They also express other markers seen on neural cells, including a neurofilament associated protein and the LA4 neuronal glycolipid (Fig. 7.3). Another marker of the neural state is the high ratio of neuropeptide CGRP relative to calcitonin hormone mRNA, as discussed below (Fig. 7.4). Furthermore, electrophysiological recordings have revealed neural-type ion channels with an ω-conotoxin sensitive N-type Ca^{+2} current and a tetrodotoxin-sensitive Na^{+} current. Finally, the CA77 C-cells have a number of serotonergic properties, as discussed below.[10,11,26,29]

Table 7.1. Neuronal properties of the CA77 C-cell line*

Morphology	
	Neurites and growth cones
	Clear and dense core vesicles
Biochemical	
	Neurofilaments (NF-L, NF-M, NF-H)
	LA4 neuronal glycolipid
	N-type calcium channels
	TTX-sensitive sodium channels
	High CGRP/calcitonin mRNA ratio
	MASH-1 and Brn-3 transcription factors
Serotonergic	
	Tryptophan hydroxylase mRNA
	5-HT_{1B} autoreceptor
	Neuronal 5-HT transporter mRNA
	SSRI-sensitive 5-HT reuptake
	Regulated 5-HT secretion

*table is modified from reference 26.

The TT cell line was established by Leong and colleagues from a human MTC.[16] While, as mentioned above, TT cells do not have an intrinsic neuronal-like morphology, they can be induced to become more neuronal. A reduced serum medium has been shown to enhance neuronal differentiation.[9] These changes appeared to be permanent, since readdition of the fetal bovine serum medium did not alter morphology. Changes in morphology were accompanied by increases in neurofilament, monoamine oxidase A (MAO A) and serotonin-binding protein (SBP) immunoreactivity. However, 5-HT immunoreactivity decreased, although this may be the consequence of increased release or decreased reuptake.[9] It should be noted that TT cells were shown to lack the trkA receptor and did not respond to NGF.

A particularly interesting aspect of normal and tumor C-cells is their serotonergic features. The pioneering work in this area was performed by Gershon and colleagues initially using primary C-cell cultures from bats and later sheep to document the synthesis of 5-HT from tryptophan,[5] and storage of 5-HT in granules bound to a neuronal binding protein (SBP).[8] More recently, serotonergic properties have also been shown using primary rat C-cell cultures and the rat CA77 and human TT cell lines. Tamir et al[9] have

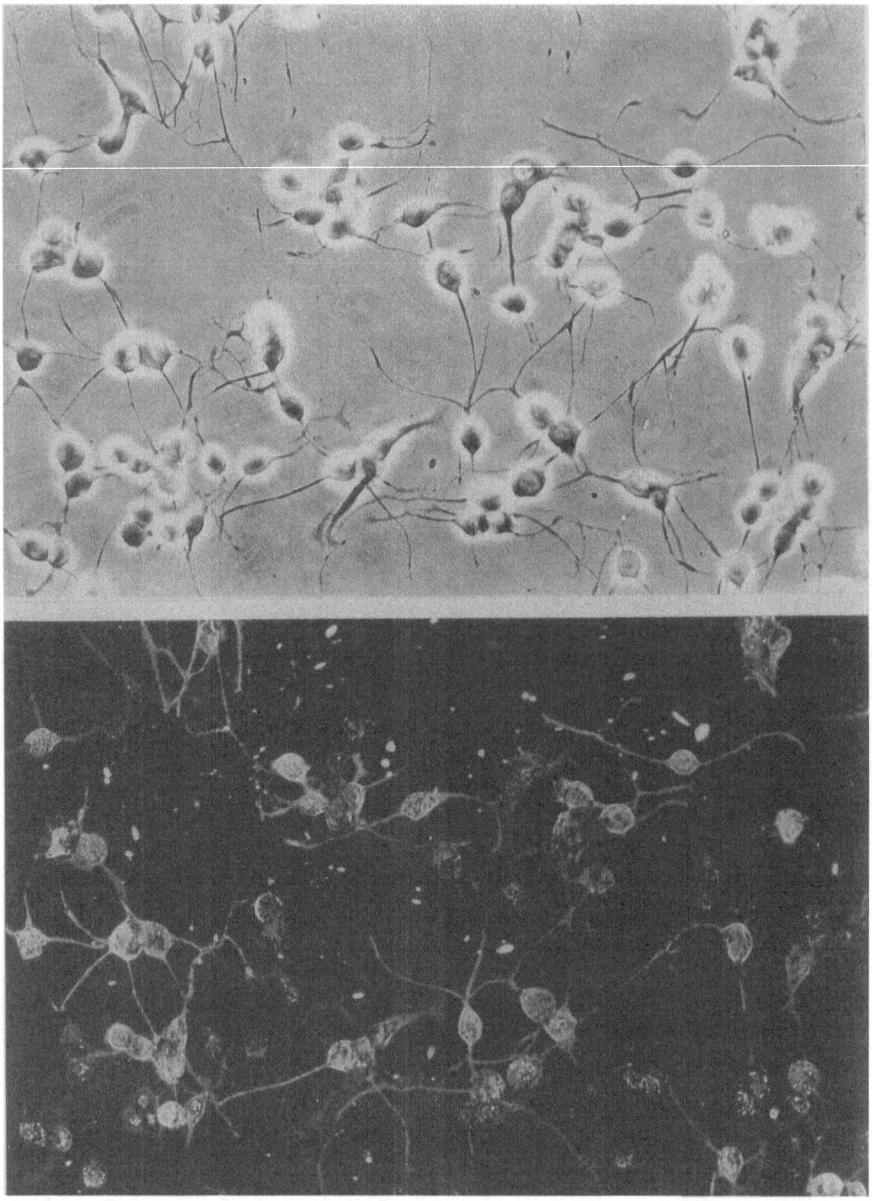

Fig. 7.2. Neurites in CA77 C-cells. (A) Phase contrast micrograph of CA77 cells with neuritic extensions. (B) Immunofluorescence micrograph of the same field stained with the LA4 antibody, which recognizes a glycolipid found on a subset of dorsal root ganglion neurons. Cells were grown on plastic dishes in serum-containing medium.

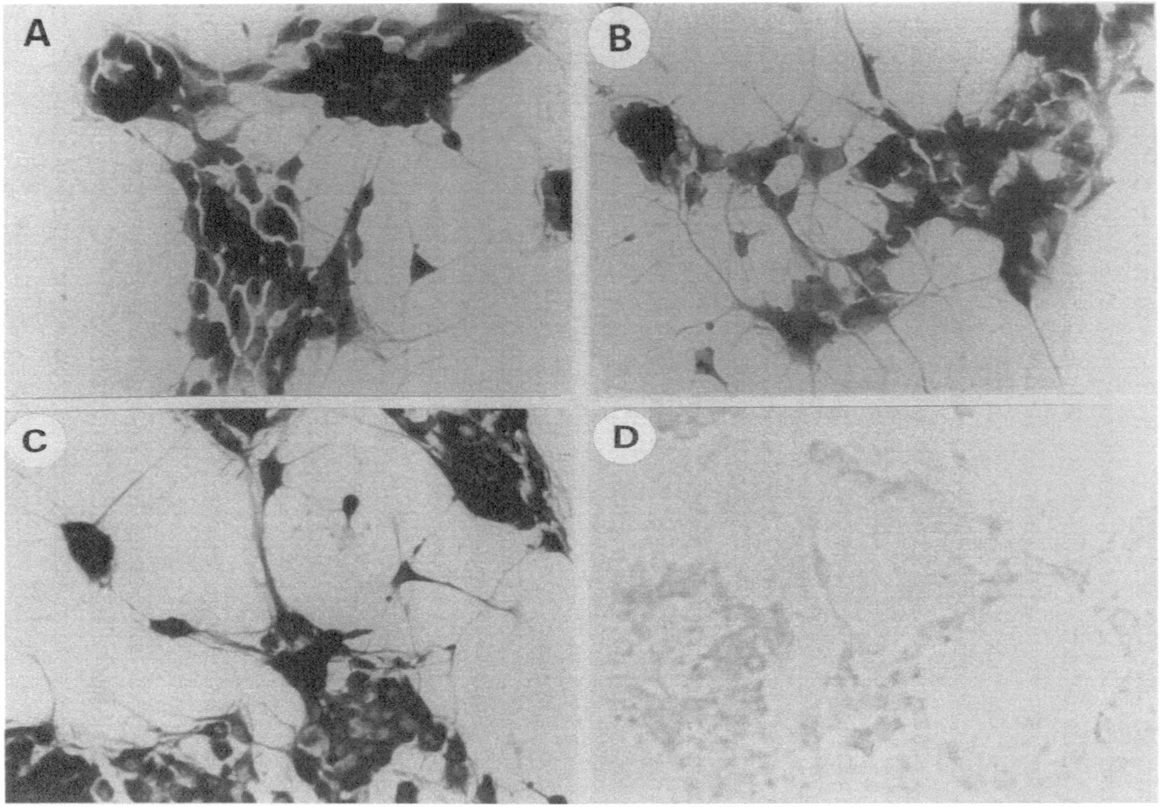

Fig. 7.3. Immunohistochemical identification of neuronal markers in CA77 cells. (A) Neurofilament staining with RT97 antibody, which recognizes NF-H. (B) Neurofilament-associated protein staining with 3A10 antibody. (C) Neuronal glycolipid staining with LA4 antibody. (D) Negative control showing lack of staining with the myeloma NS1 supernatant in place of the primary antibody. Descriptions of the monoclonal antibodies and immunostaining procedure are described in Russo et al.[10] Briefly, cells were fixed in methanol:acetone (1:1), treated with 0.017% Triton X-100 and 0.75% gelatin, and then incubated overnight at 4 °C with antibodies. Immunostaining was detected by peroxidase-coupled staining using diaminobenzidine with nickel and cobalt enhancement.

reported that the TT MTC cell line synthesizes 5-HT and also possesses a fluoxetine inhibitable reuptake mechanism. In addition, TT cells contain MAO A, 45 kDa and 56 kDa SBPs that are apparently identical to those in the CNS,and 5-HT storage vesicles with a reserpine sensitive biogenic amine transporter.[9] Finally, the TT cells show regulated secretion of 5-HT in response to extracellular calcium and thyroid stimulating hormone.[30] Interestingly, 5-HT release from C-cells may form a paracrine regulatory loop controlling thyroid hormone release from the adjacent follicular cells.[31] Many of these properties are associated with serotonergic neurons, but not with non-neuroectodermal cells that store 5-HT (platelets, mast cells). The CA77 cell line has also recently been shown to possess serotonergic properties,[11,26,29] including fluoxetine-inhibitable 5-HT uptake, regulated release of 5-HT, the 5-HT_{1B} autoreceptor and the presence of tryptophan hydroxylase (the rate limiting enzyme in 5-HT synthesis) mRNA (Table 7.1). The presence of serotonin synthesis, regulated secretion, uptake and feedback control constitute the fundamental features of serotonergic neurons. Hence, these observations suggest that the CA77 and TT C-cell lines may be of considerable utility to investigators studying serotonergic cellular and molecular mechanisms.

One of the clinical markers of MEN 2 is overproduction of the hormone calcitonin from MTC. Calcitonin is produced by alternative splicing of the calcitonin/CGRP transcript, and is a marker of the endocrine C-cell phenotype.[32] In contrast to C-cells, neurons express the neuropeptide CGRP as the predominant splicing product from the calcitonin/CGRP gene.[33] CGRP has been assigned a number of functions, with vasodilation of peripheral and cerebral arterioles being the best characterized.[34,35] Consistent with the above mentioned neuronal transdifferentiation, there is a shift from calcitonin to CGRP production in C-cell tumors upon serial passage.[36] While normal C-cells express less than 5% of calcitonin/CGRP transcripts as CGRP,[32] the CA77 C-cell line, for example, expresses 80-90% of the transcripts as CGRP (Fig. 7.4).[37] In humans, MTC has also been shown to progress from high to low calcitonin production states, which also serves as a clinical indicator of tumor progression (chapter 8). This shift can also be replicated by NGF treatment of primary C-cells, which then show increased production of CGRP.[8]

While the neuronal features of the C-cell tumor lines are interesting with respect to the consequences of MEN 2, a pertinent

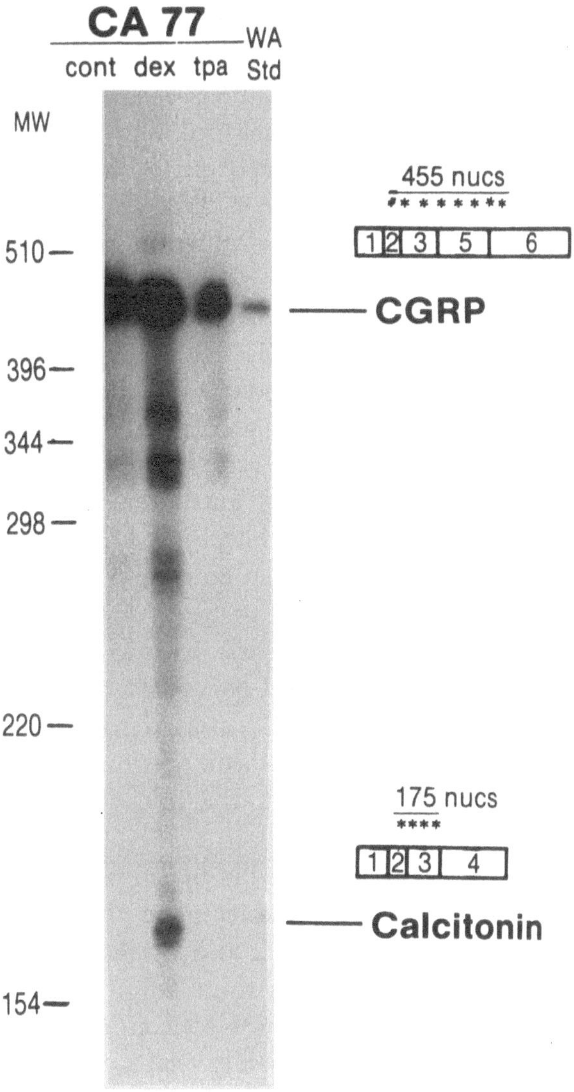

Fig. 7.4. Simultaneous measurement of calcitonin and CGRP RNA levels in CA77 cells before and after dexamethasone treatment. A ribonuclease protection assay was performed using total RNA (20 μg) isolated from CA77 cells treated with 0.1 μM dexamethasone (dex), 0.004% ethanol (cont), or 0.1 uM phorbol 12-myristate 13-acetate (TPA) for 6 days. The CGRP cRNA probe and the protected species following hybridization to CGRP mRNA (455 nucleotides) and calcitonin mRNA (175 nucleotides) are indicated. As a control, poly(A) enriched RNA (0.1 μg) from the WA MTC tumor was used since it contains equivalent amounts of calcitonin and CGRP mRNAs. Size standards are pBR322 digested with HinfI.

question is whether this transdifferentiation is a nuance of tumor cells or could it occur in normal C-cells? This point was resolved by showing that primary C-cell cultures become more neuronal when treated with NGF.[8,11] These enhanced changes include greater neurite extension, increased expression of neurofilaments and CGRP and increased serotonin transporter mRNA and fluoxetine-sensitive uptake activity. In addition, co-culturing with aneural chick hindgut,[8] and plating on laminin substratum,[38] have also been shown to enhance neuronal morphology of primary C-cells. Why should C-cells be responsive to NGF if they do not normally differentiate as neurons in vivo? One speculation, discussed by Barasch et al,[8] is that NGF is needed as a trophic agent for C-cell survival. This possibility is supported by the generation of hypothyroidism in rats administered NGF antiserum.[39] Further support for a trophic role is the observation that NGF increased DNA synthesis and proliferation of human MTC cells in culture.[40] Alternatively, the retention of NGF responsiveness may be a developmental holdover reflecting the ontogeny of C-cells from a sympathoadrenal neural crest progenitor. This would be consistent with the well-documented role of NGF as a neurotrophic agent for sympathoadrenal crest-derived sympathetic neurons.[12] In either case, the NGF actions provide insight into the role of tumorigenesis in the C-cell phenotype.

How can tumorigenesis enhance the neuronal phenotype? The key to that question may lie in the ret gene that is activated in MEN 2. As discussed in Chapters 2 and 3, ret is a receptor tyrosine kinase that transmits signals from an extracellular ligand, and it has been genetically linked to C-cell tumors.[17-20] This circumstantial evidence raises the possibility that ret activity may also be responsible for the neuronal properties common in cell lines derived from these tumors. Yet mechanistically does that make sense? First, the significance of ret in neuronal differentiation has been clearly demonstrated by studies on mice lacking the ret gene.[23] While enteric neuroblasts form and migrate in these mice, the cells fail to fully differentiate into mature neurons. In normal mice, the ret ligand would presumably activate neuronal differentiation in the gut, while C-cells in the thyroid gland would not encounter the putative ret ligand and hence would not realize their neuronal potential. Consequently, the activated ret receptor could induce not only hyperplasia but also the dormant neuronal pathway in

C-cells in MEN 2. While tyrosine kinase signals are generally associated with cell growth, there is precedence for such signals to cause neuronal differentiation. The neurotrophins NGF, NT-3, NT-4/5 and BDNF all act through the trk tyrosine receptor family.[41] Perhaps the best-characterized example is NGF induction of neuronal differentiation with the PC12 pheochromocytoma cell line (see Chapter 4). Likewise, NGF also enhances the neuronal properties of C-cells. Consequently, an attractive model is that both NGF and ret may be activating similar or overlapping signaling pathways leading to neuronal differentiation (Fig. 7.1).

GLUCOCORTICOID AND RAS REPRESSION OF NEURONAL PROPERTIES AND CELL GROWTH

Neuronal features in CA77 cells can be repressed by glucocorticoids.[10] Treatment with the synthetic glucocorticoid, dexamethasone, for several days induced thinning and partial retraction of neurites, as well as increased cell roundness (Fig. 7.5). In addition, dexamethasone treatment increased the number of secretory granules by 2-fold, increased calcitonin mRNA levels 3-4-fold relative to CGRP mRNA (Fig. 7.4) and decreased neurofilament NF-L mRNA levels. All effects were reversible upon removal of dexamethasone.[10] These features are consistent with the finding that dexamethasone also induces cell rounding and biased production of calcitonin mRNA in the TT MTC cell line.[10,42] These observations suggest that an endocrine differentiation pathway has been induced by dexamethasone. Perhaps most striking among the dexamethasone-induced changes was a 10-fold reduction in DNA synthesis with a similar decrease in cell number.[10]

Retinoic acid also induced morphological effects on the CA77 cells.[10] There was a marked retraction of neurites and cell rounding with reduced cell adhesion. There was also an increase in the number of cytoplasmic secretory granules. However, retinoic acid did not reduce DNA synthesis, alter the calcitonin to CGRP mRNA ratio, or decrease neurofilament NF-L expression. Thus, the retinoic acid effects are primarily morphological and appear to be the consequences of reduced cell adhesion. It is interesting that retinoic acid also reduces neural crest cell adhesion with adverse effects on cell migration.[43]

It is particularly interesting that similar changes can be induced by infection of TT cells with the Harvey murine sarcoma virus.[44]

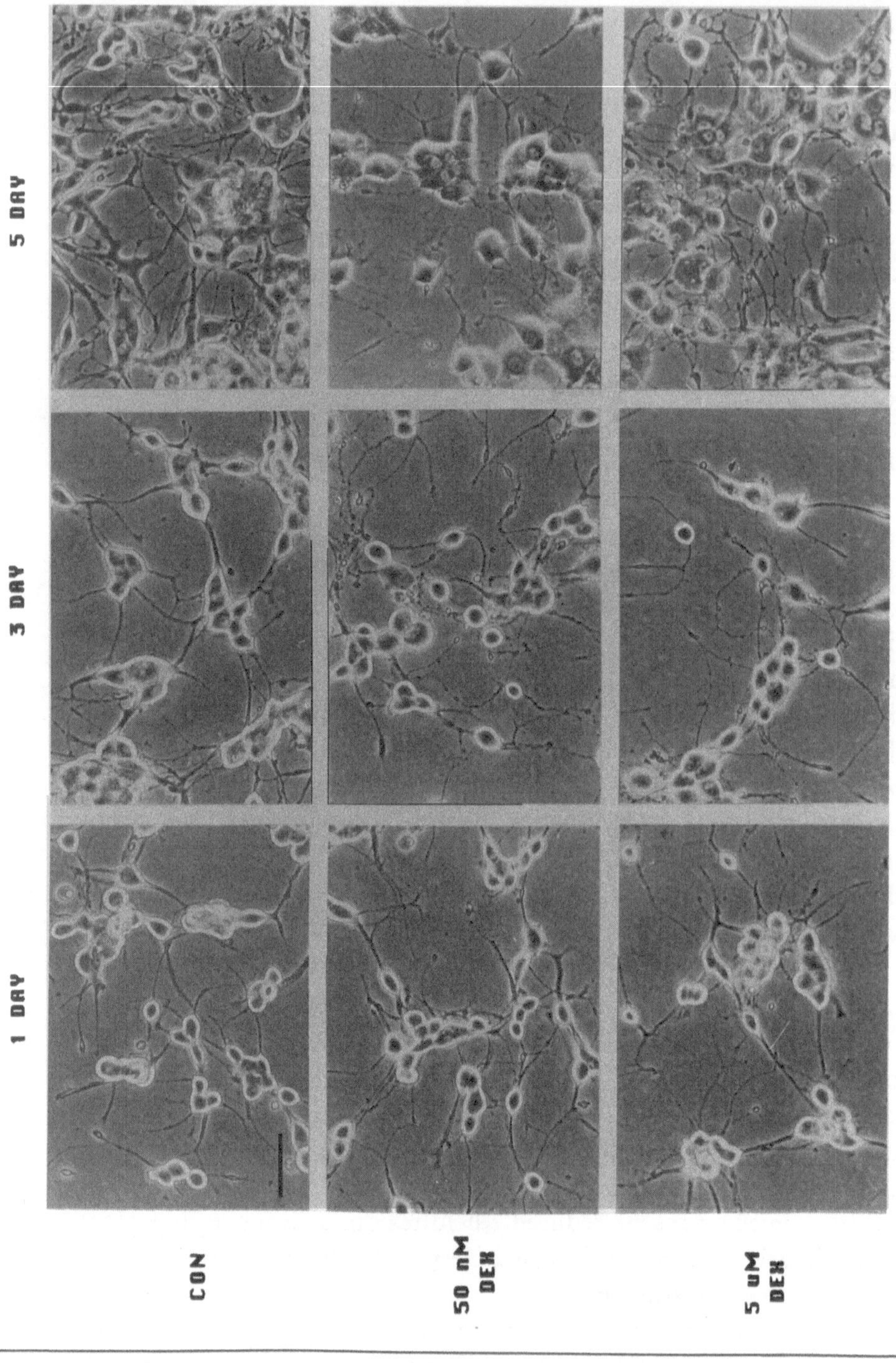

Fig. 7.5. Retraction and thinning of neurites in CA77 cells induced by dexamethasone. CA77 cells were plated in serum-free medium on laminin-coated dishes and treated for 1, 3, or 5 days with the ethanol vehicle (CON), or 50nM, 0.5 μM, or 5 μM dexamethasone (DEX), as indicated.

This virus expresses the Harvey ras oncogene, which is a constitutively active form of the cellular ras signal transduction protein. The ras-induced changes include a more rounded morphology, decreased thymidine incorporation rate, increased secretory vesicles and calcitonin secretion and increased calcitonin to CGRP mRNA ratio. It is somewhat surprising that similar properties can be induced by two such different agents. These phenotypic similarities suggest the possibility that dexamethasone and ras may act on a common target in the C-cells.

How might dexamethasone and ras repress the neuronal phenotype? An intriguing possibility is that they may be down-regulating expression of the ret gene. The possibility that dexamethasone inhibits ret expression is currently being tested in our lab. While speculative, this hypothesis suggests that glucocorticoid treatment may be useful as a nonsurgical option for medullary thyroid carcinoma. An alternative non-exclusive hypothesis is that dexamethasone and ras might somehow be interfering with the ret signaling pathway. This is especially reasonable for ras since it is a central component in tyrosine kinase signaling pathways and there is considerable cross-talk and overlap between these paths (e.g., ref. 45). In both cases, the decreased cell growth caused by dexamethasone and ras might then be key to the other phenotypic changes, since Nelkin and colleagues have observed that reduced growth in TT cells leads to an increase in the calcitonin/CGRP ratio.[46]

IDENTIFICATION OF DIFFERENTIATION GENES BY DIFFERENTIAL cDNA SCREENING

It is also possible to use C-cell lines to identify genes regulated by changes in differentiation states. Differential screening is a powerful approach for identifying genes involved in cell differentiation and/or growth.[47] We have performed a PCR-based plus/minus differential screening of a CA77 cDNA library using cDNA probes prepared from control and dexamethasone-treated CA77 cells.[48,49] Likewise, Helman et al[50] have used differential screening to identify specific genes from MTC tumors.

We have identified a novel isoform of $G_s\alpha$, termed $G_s\alpha N1$, that is induced by dexamethasone (Fig. 7.6).[48] $G_s\alpha N1$ is an alternatively spliced and polyadenylated variant that contains only the N-terminal 86 amino acids of $G_s\alpha$ followed by a previously

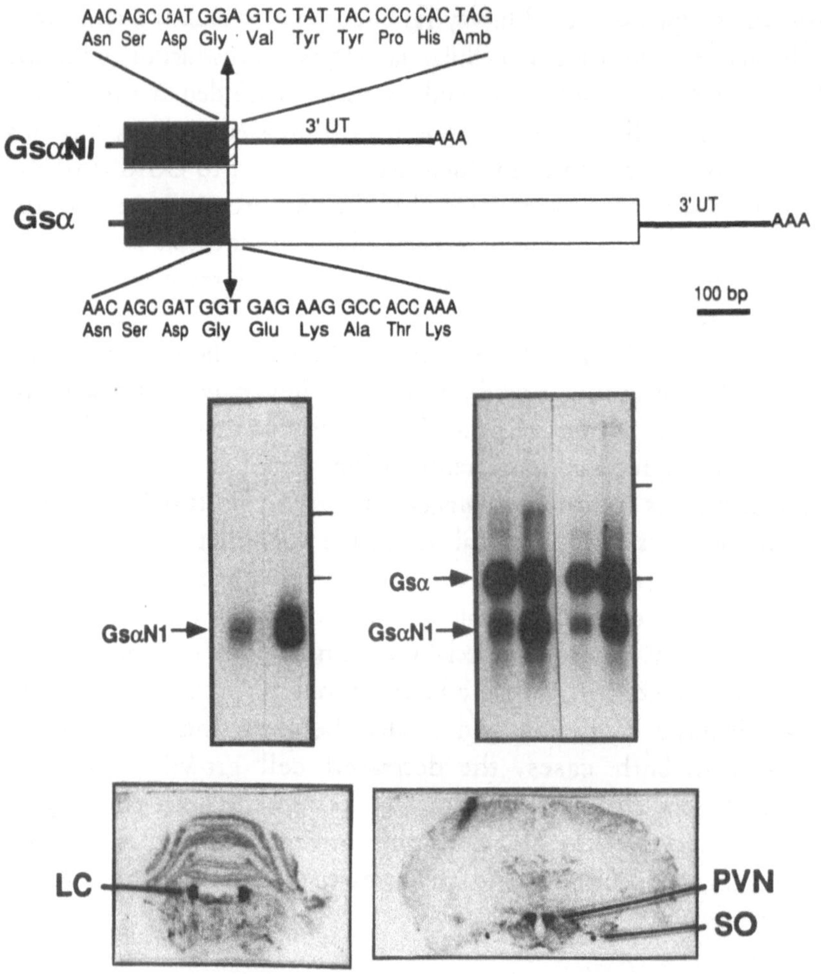

Fig. 7.6. Identification of a dexamethasone-induced novel truncated G protein mRNA. The top portion shows a schematic representation of $G_s\alpha N1$ compared to $G_s\alpha$. The filled boxes represent identical sequences. The alternative splice site, which introduces new sequences into $G_s\alpha N1$, is shown by the arrows, with the flanking sequences given for both $G_s\alpha$ and $G_s\alpha N1$. The middle panels are Northern blot autoradiograms showing dexamethasone induction of $G_s\alpha N1$ mRNA. The left panel is following hybridization with a $G_s\alpha N1$ specific probe. The right panel is following hybridization with a probe that recognizes both $G_s\alpha N1$ and $G_s\alpha$ mRNAs. Hatch marks indicate the positions of 18S and 28S ribosomal RNA. The lower panels show representative in situ hybridization autoradiograms of $G_s\alpha N1$ expression sites. Coronal rat brain sections show expression in the locus coerleus (LC) of the brainstem (left), and in the hypothalamic paraventricular nucleus (PVN) and supraoptic nucleus (SO) (right). Figure is modified from reference 48.

undescribed exon encoding only 5 additional amino acids. The most interesting feature of $G_s\alpha N1$ is its restricted localization in comparison to the wide expression pattern of $G_s\alpha$.[48] Northern and in situ hybridizations have shown that $G_s\alpha N1$ is primarily located in the brain and brainstem, with lower levels in the thyroid, adrenal medulla and skeletal muscle (Fig. 7.6). Within the brainstem, it is particularly abundant in nuclei containing biogenic amines—noradrenergic neurons in the locus coeruleus and serotonergic raphe neurons. In the hypothalamus, expression is especially high in peptidergic neurons, such as the paraventricular nucleus. In the adrenal gland, expression is restricted to just the chromaffin cells in the medulla.

While the function of $G_s\alpha N1$ is not yet known, one speculation based on its expression pattern is that it is involved in regulated secretion from neurons and neuroendocrine cells.[48] In this regard, it is interesting that injection of antibodies directed against the amino-terminal region of $G_s\alpha$ inhibited trans-Golgi vesicular transport in epithelial cells, while C-terminal antibodies were ineffective.[51] Recently, another alternative form of $G_s\alpha$, termed $XLG_s\alpha$, was found to contain additional N-terminal sequences and was localized to many of the same neuronal and endocrine sites as $G_s\alpha N1$.[52] These studies emphasize the potential importance of the N-terminal region contained in $G_s\alpha N1$.

An alternative, but not exclusive, hypothesis is that $G_s\alpha N1$ may act as a growth suppressor. As mentioned above, one of the most pronounced effects of dexamethasone treatment on the CA77 cells is to arrest cell growth. G proteins are well-positioned as central controllers of cellular functions.[53] How might $G_s\alpha N1$ be involved in growth regulation in the MTC cells? Previous studies on $G\alpha$ proteins have assigned different functions to regional domains of $G_s\alpha$.[54,55] These structure-function studies predict that $G_s\alpha N1$ itself will be unable to transmit a signal directly, but may be capable of binding $\beta\gamma$-subunits and act as a scavenger by sequestering free $\beta\gamma$-subunits. Since it has now been established that the $\beta\gamma$-subunits may also transmit signals,[56] then signals sent by $\beta\gamma$-subunits would be similarly affected. In this role, $G_s\alpha N1$ might attenuate growth factor signals transmitted by G proteins in MTC cells. $G_s\alpha N1$ might also bind proteins other than $\beta\gamma$ subunits, as seen with $G\alpha$-subunits.[57] For example, $G\alpha$ appears to interact with a proto-oncogene ras-GAP complex, which might influence the effects of ras on cell growth and differentiation. In this regard, it is intriguing that the activated ras oncogene can cause differentiation of TT cells in a similar manner to dexamethasone treatment, as discussed above. We are currently investigating the binding properties and function of $G_s\alpha N1$. It is interesting to note that there is precedence for the involvement of G proteins in other growth and differentiation systems,[57-59] including endocrine tumors of the pituitary.[60]

TRANSCRIPTIONAL REGULATION OF THE CALCITONIN/CGRP GENE

Cell lines are particularly useful in the analysis of cis- and trans-acting transcription factors, due to the homogeneity of cell type and the ease of transient transfection assays with reporter genes. The C-cells are an especially versatile system for these studies since they express a relatively large number of neuropeptides, including calcitonin, CGRP, somatostatin, NPY, CCK and substance P, that are expressed in C-cells (unpublished data),[5,61,62,63] similar to the host of peptides expressed in enteric neurons.[64,65] Transcriptional studies, while applicable to any gene expressed in MEN 2, will be illustrated for regulation of the calcitonin/CGRP gene.

The calcitonin/CGRP gene, as described above, is transcribed in both neural and endocrine cell types. It is regulated by a variety of agents including protein kinases A and C,[66] vitamin D,[67,68] retinoic acid[69] and glucocorticoids.[37,70] In order to investigate the cell-specific and hormonal regulation, we fused 5' flanking DNA from the CT/CGRP gene to luciferase and chloramphenicol acetyl transferase reporter genes. These were transfected into CA77 cells (and another C-cell line, 44-2C) and an 18 bp region was identified as a cell-specific enhancer by progressive deletions and mutations (Fig. 7.7).[70,71] Helix-loop-helix (HLH) and octamer (Oct) transcription factor motifs were identified within this sequence, and the functional significance of these elements was proven by point mutations that greatly reduced activity. A similar enhancer sequence has been shown to be important for cell-specific expression of the human calcitonin/CGRP gene.[72-74]

The DNA binding proteins that recognized the HLH and octamer motifs were characterized by electrophoretic mobility shift assays. Cell-specific complexes were identified by using radiolabeled 18 bp calcitonin/CGRP enhancer DNA incubated with CA77 (or

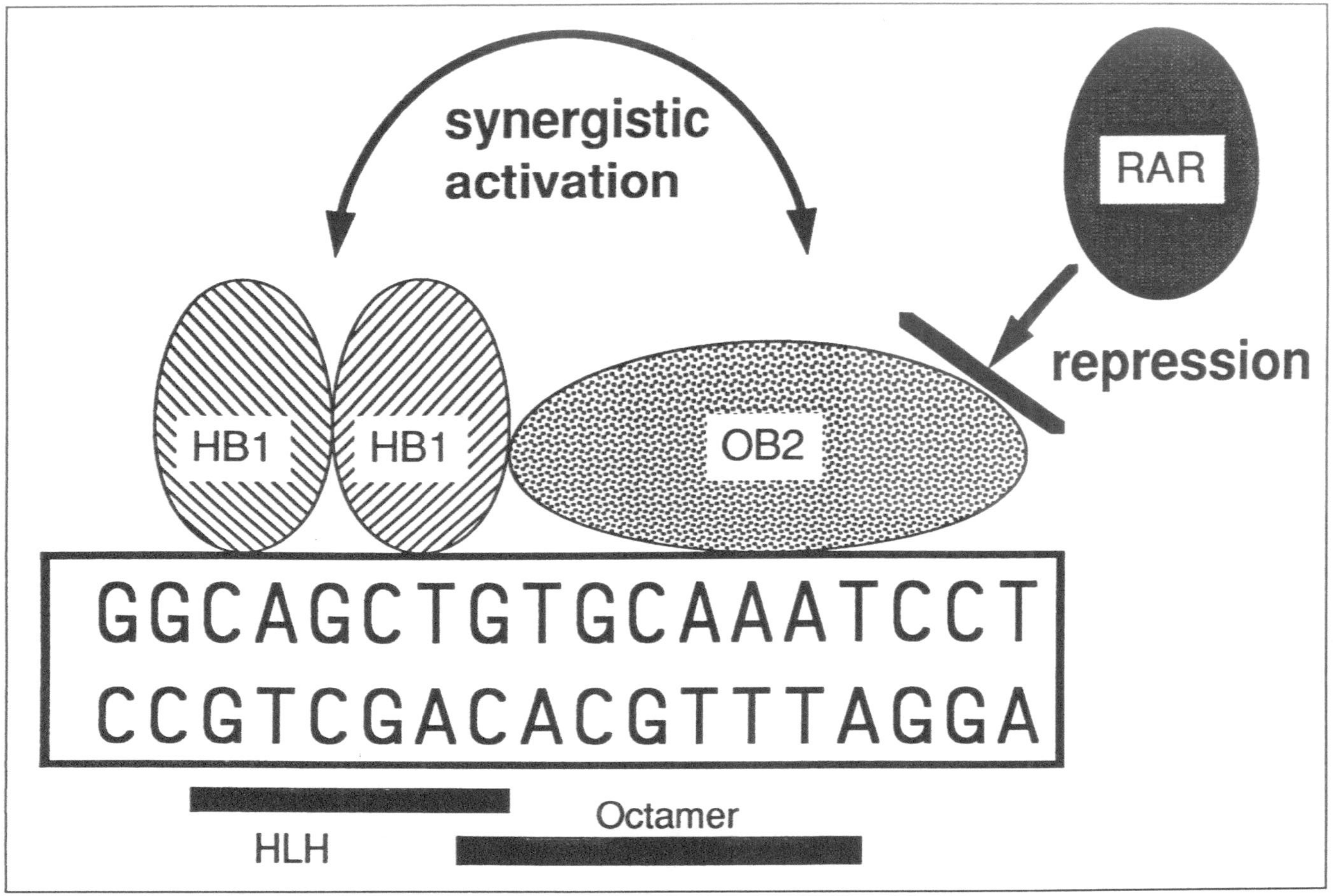

Fig. 7.7. Activation and repression of the calcitonin/CGRP enhancer. The 18 bp rat calcitonin/CGRP enhancer sequence is shown (position -1043 to -1025). The HLH and octamer motifs and corresponding binding proteins are indicated. HB1 is the HLH dimer complex and OB2 is the cell-specific octamer binding protein. Activation of the enhancer requires synergistic contributions from both of these factors. The enhancer is repressed by retinoic acid receptor monomers (RAR). RAR does not bind the DNA element, but rather, inhibits through protein-protein interactions.

44-2C) nuclear extracts, and compared to complexes seen with HeLa nuclear extracts. HeLa cells do not express the calcitonin/ CGRP gene and hence served as a negative control. The specific complexes were further characterized by competition with unlabeled oligonucleotides containing consensus HLH and octamer motifs.[71] Two cell-specific bands were identified, HB1 corresponding to an HLH protein and OB2 corresponding to an octamer factor. In addition, binding of the ubiquitous Oct-1 protein was documented by antibody supershift assays. We also used antiserum directed against the MASH-1 HLH protein to show that MASH-1 can bind the calcitonin/CGRP enhancer (unpublished data). In agreement with this binding data, cotransfection of a MASH-1 expression vector yielded a small, but significant, increase in the enhancer activity.[71] Also, the human MASH-1 protein (called hASH-1) has been cloned and shown to be expressed in the TT C-cell line.[75]

A fundamental feature of the calcitonin/CGRP enhancer is that it involves synergism between the HLH and octamer motifs for full activity (Fig. 7.7).[71] The presence of only one motif yields less than 2-4-fold activation, in contrast to the 20-30-fold activation seen with both motifs. Likewise, when either site was mutated or separated by the insertion of 1 bp, 5 bp or 10 bp, the activity was greatly reduced (unpublished data).[71] Synergism between transcription factors is an emerging theme that will probably be the rule more than the exception.[76] Such a combinatorial code consisting of multiple factors allows a smaller number of factors to control the estimated 100,000 genes in the human genome. The abundance of octamer and HLH factors suggests that interactions between these classes of proteins will be a common feature in gene regulation.

Interestingly, this same enhancer was also shown to be responsible for glucocorticoid[70] and retinoic acid[69] repression of calcitonin/CGRP transcription (Fig. 7.7). Nuclear receptors repress transcription by several mechanisms.[77] We have shown that retinoic acid repression of the calcitonin/CGRP gene is mediated by protein-protein interactions.[69] This observation is supported by a report that the DNA binding domains of glucocorticoid receptors and Oct-1 can interact.[78] The potential interaction between RAR and octamer proteins is especially interesting, since it suggests that retinoids may exert some of their known anti-proliferative actions on cancer cells[79] via inhibitory interactions with octamer proteins involved in cellular proliferation.

These studies have provided the foundation to not only examine calcitonin/CGRP transcription, but have also provided some insight into the broader question of C-cell development. Our studies on the calcitonin/CGRP enhancer have focused our attention on MASH-1. MASH-1 is a transcription factor that has been shown to be present in neuroblasts, but not in terminally differentiated neurons.[80,81] The importance of MASH-1 in neurogenesis has been proven, using homozygous MASH-1 deletion mice. These mice are non-viable and show severe defects in autonomic ganglia, in which neuronal precursors are present but differentiation does not proceed.[82] We have shown that MASH-1 RNA is present in the juvenile rat thyroid.[11] At present, the only other example of MASH-1 expression in the postnatal animal is in the olfactory neuroblasts. Detection of MASH-1 in C-cells suggests that C-cells may retain the neurogenic potential of sympathoadrenal progenitors. Furthermore, the subsequent loss of MASH-1 in the primary cultures is consistent with neuronal differentiation.[11] These results support the hypothesis, discussed above, that the thyroid environment may repress neuronal features in C-cells.

SUMMARY

Thyroid C-cells share a common ontogeny with enteric serotonergic neurons and can be induced to express a neuronal phenotype following NGF treatment. C-cell lines that have been established from medullary thyroid carcinomas of MEN 2 share these neuronal features. We have described the neuronal properties of the CA77 and TT C-cell lines and proposed that these properties are the result of the activated ret kinase activity, analogous to the actions of ret in enteric neurogenesis. The CA77 and TT cell lines show reversible endocrine/neuronal differentiation and hence provide a system for identifying genes important in growth control and phenotypic specification, as exemplified by the $G_s\alpha N1$ transcript. Finally, the C-cell lines provide an amenable system to examine transcriptional regulation of genes expressed in MEN 2 tumors, for example, neuroendocrine peptides and neurotransmitters. Studies on the neuronal properties of thyroid C-cell lines should further our understanding of the molecular control mechanisms in normal differentiation and pathogenesis of MEN 2.

ACKNOWLEDGMENTS

We gratefully acknowledge our colleagues in the C-cell and calcitonin/CGRP field for many stimulating discussions. This work was supported by NIH grant HD 25969 to A.R., with tissue culture support provided by the Diabetes and Endocrinology Center (DK 25295) and antibodies supplied by the Developmental Studies Hybridoma Bank (HD 23144).

REFERENCES

1. Austin LA, Heath H. Calcitonin physiology and pathophysiology. N Engl J Med 1981; 304:269-275.
2. Copp DH. Remembrance: Calcitonin: Discovery and early development. Endocrinol 1992; 131:1007-1008.
3. Brown EM. Extracellular Ca^{2+} sensing, regulation of parathyroid cell function, and role of Ca^{2+} and other ions as extracellular (first) messengers. Physiol Rev 1991; 71:371-411.
4. McDermott MT, Kidd GS. The role of calcitonin in the development and treatment of osteoporosis. Endo Rev 1987; 8:377-390.
5. Nunez EA, Gershon MD. Synthesis and storage of serotonin by parafollicular (C) cells of the thyroid gland of active, prehibernating and hibernating bats. Endo 1972; 90:1008-1024.
6. Barasch JM, Tamir H, Nunez EA et al. Serotonin-storing secretory granules from thyroid parafollicular cells. J Neurosci 1987; 7:4017-4033.
7. LeDouarin NM. The Neural Crest. Cambridge: Cambridge University Press, 1982.
8. Barasch JM, Mackey H, Tamir H et al. Induction of a neural phenotype in a serotonergic endocrine cell derived from the neural crest. J Neurosci 1987; 7:2874-2883.
9. Tamir H, Liu K, Payette RF et al. Human medullary thyroid carcinoma: characterization of the serotonergic and neuronal properties of a neurectodermally derived cell line. J Neurosci 1989; 9:1199-1212.
10. Russo AF, Lanigan TM, Sullivan BE. Neuronal properties of a thyroid C-cell line: repression by dexamethasone and retinoic acid. Mol Endo 1992; 6:207-218.
11. Clark MS, Lanigan TM, Page N et al. Induction of a serotonergic and neuronal phenotype in thyroid C-cells J Neurosci 1995; 15:6167-6178.
12. Anderson DJ. Molecular control of cell fate in the neural crest: the sympathoadrenal lineage. Annu Rev Neurosci 1993; 16:129-158.

13. Jacobs-Cohen RJ, Tamir H, Gershon MD. Expression of a neuronal phenotype by neural crest-derived paraneurons (parafollicular cells) is antagonized by thyroid hormone (triiodothyronine; T3). Soc Neurosci Abstracts 1994; 20:654.
14. Lendahl U, McKay RDG. The use of cell lines in neurobiology. Trends Neurosci 1990; 13:132-137.
15. Muszynski M, Birnbaum RS, Roos BA. Glucocorticoids stimulate the production of preprocalcitonin-derived secretory peptides by a rat medullary thyroid carcinoma cell line. J Biol Chem 1983; 258:11678-11683.
16. Leong SS, Horoszewicz JS, Shimaoka K et al. In: Andreoli M, Manaco F, Robbins J eds. Advances in Thyroid Neoplasia. Rome: Field Educational Italia, 1981; 95-108.
17. Santoro M, Rosati R, Grieco M et al. The ret proto-oncogene is consistently expressed in human pheochromocytomas and thyroid medullary carcinomas. Oncogene 1990; 5:1595-1598.
18. Mulligan LM, Eng C, Healey CS et al. Specific mutations of the RET proto-oncogene are related to disease phenotype in MEN 2A and FMTC. Nature Genet 1994, 6:70-74.
19. Hofstra RMW, Landsvater RM, Ceccherini I et al. A mutation in the RET proto-oncogene associated with multiple endocrine neoplasia type 2B and sporadic medullary thyroid carcinoma. Nature 1994; 367:375-376.
20. Santoro M, Carlomagno F, Romano A et al. Activation of RET as a dominant transforming gene by germline mutations of MEN2A and MEN2B. Science 1995; 267:381-383.
21. Romeo G, Ronchetto P, Luo Y et al. Point mutations affecting the tyrosine kinase domain of the RET proto-oncogene in Hirschsprung's disease. Nature 1994; 367:377-378.
22. Edery P, Lyonnete S, Mulligan LM et al. Mutations of the RET proto-oncogene in Hirschsprung's disease. Nature 1994; 367:378-380.
23. Schuchardt A, D'Agati V, Larsson-Blomberg L et al. Defects in the kidney and enteric nervous ystem of mice lacking the tryrosine kinase receptor Ret. Nature 1994; 367:360-383.
24. Schulz N, Propst F, Rosenberg MP et al. Pheochromocytomas and C-cell thyroid neoplasms in transgenic c-mos mice: a model for the human multiple endocrine neoplasia type 2 syndrome. Cancer Res 1992; 52:450-455.
25. Baetscher M, Schmidt E, Shimizu A et al. SV40 T antigen transforms calcitonin cells of the thyroid but not CGRP-containing neurons in transgenic mice. Oncogene 1991; 7, 1133-1138.
26. Clark MS, Lanigan TM, Russo AF. Serotonergic neuronal properties in C-cell lines. Methods: A Companion to Methods Enzymol 1995; 7:253-261.
27. Wiedenmann B, Franke WW, Kuhn et al. Synaptophysin: a marker protein for neuroendocrine cells and neoplasms. Proc Natl Acad

Sci USA 1986; 83:3500-3504.
28. Sikri KL, Varndell IM, Hamid QA et al. Medullary carcinoma of the thyroid. Cancer 1985; 56:2481-2491.
29. Clark MS, Johnson W, Russo AF. Dexamethasone repression of tryptophan hydroxylase mRNA levels in the CA77 cell line. Soc Neurosci Abstracts 1994; 20:289.
30. Tamir H, Liu K, Hsiung S et al. Multiple signals leading to the secretion of 5-hydroxytryptamine by MTC cells, a neuroectodermally dervived cell line. J Neurosci 1990; 10:3743-3753.
31. Tamir H, Hsiung S, Yu P et al. Serotonergic signalling between thyroid cells: protein kinase C and 5-HT2 receptors in the secretion and action of serotonin. Synapse 1992; 12:155-168.
32. Amara SG, Jonas V, Rosenfeld MG et al. Alternative RNA processing in calcitonin gene expression generates mRNAs encoding different polypeptide products. Nature 1982; 298:240-244.
33. Rosenfeld MG, Mermod J, Amara SG et al. Production of a novel neuropeptide encoded by the calcitonin gene via tissue-specific RNA processing. Nature 1983; 304:129-135.
34. Brain SD, Williams TJ, Tippins JR et al. Calcitonin gene-related peptide is a potent vasodilator. Nature 1985; 313:54-56.
35. Marshall I. Mechanism of vascular relaxation by the calcitonin gene-related peptide. Annals New York Acad Sci 1992; 657:204-215.
36. Rosenfeld MG, Amara SG, Roos BA et al. Altered expression of the calcitonin gene associated with RNA polymorphism. Nature 1981; 290:63-65.
37. Russo AF, Nelson C, Roos BA et al. Differential regulation of the coexpressed calcitonin/α-CGRP and β-CGRP neuroendocrine genes. J Biol Chem 1988; 263:5-8.
38. Nishiyama I, Fujii T. Laminin-induced process outgrowth from isolated fetal rat C-cells. Exp Cell Res 1992; 198:214-220.
39. Levi-Montalcini, R. Developmental neurobiology and the natural history of nerve growth factor. Annu Rev Neurosci 1982; 5:341-362.
40. Goretzki PE, Wahl RA, Becker R et al. Nerve growth factor (NGF) sensitizes human medullary thyroid carcinoma (hMTC) cells for cytostatic therapy in vitro. Surgery 1987; 102:1035-1042.
41. Chao MV. Neurotrophin receptors: a window into neuronal differentation. Neuron 1992; 9:583-593.
42. Cote GJ, Gagel RF. Dexamethasone differentially affects the levels of calcitonin and calcitonin gene-related peptide mRNAs expressed in a human medullary thyroid carcinoma cell line. J Biol Chem 1986; 261:15524-15528.
43. Thorogood P, Smith L, Nicol A et al. Effects of vitamin A on the behaviour of migratory neural crest cells in vitro. J Cell Sci 1982; 57:331-350.
44. Nakagawa T, Mabry M, deBustros A et al. Introduction of v-Ha-ras oncogene induces differentiation of cultured human medullary

thyroid carcinoma cells. Proc Natl Acad Sci USA 1987; 84: 5923-5927.
45. Ginty DD, Bonni A, Greenberg ME. Nerve growth factor activates a ras-dependent protein kinase that stimulates c-fos transcription via phosphorylation of CREB. Cell 1994; 77:713-725.
46. Nelkin BD, Chen KY, deBustros A et al. Changes in calcitonin gene RNA processing during growth of a human medullary thyroid carcinoma cell line. Cancer Res 1989; 49:6949-6952.
47. Sargent TD. Isolation of differentially expressed genes. Methods Enzymol 1987; 152:423-432.
48. Crawford JA, Mutchler KJ, Sullivan BE et al. Neural expression of a novel alternatively spliced and polyadenylated $G_s\alpha$ transcript. J Biol Chem 1993; 268:9879-9885.
49. Mutchler KJ, Klemish SW, Russo AF. A rapid PCR protocol for identification of differentially expressed genes from a cDNA library. PCR Methods and Applic 1992; 1:195-198.
50. Helman LJ, Thiele CJ, Linehan WM et al. Molecular markers of neuroendocrine development and evidence of environmental regulation. Proc Natl Acad Sci USA 1987; 84:2336-2339.
51. Pimplikar SW, Simons K. Regulation of apical transport in epithelial cells by a G_s class of heterotrimeric G protein. Nature 1993; 362:456-458.
52. Kehienbach RH, Matthey J, Huttner WB. XLαs is new type of G protein. Nature 1994; 372:804-809.
53. Spiegel AM. G proteins in cellular control. Curr Opin Cell Biol 1992; 4:203-211.
54. Conklin BR, Bourne HR. Structural elements of Gα-subunits that interact with Gβγ, receptors, and effectors. Cell 1993; 73:631-641.
55. Johnson GL, Dhanasekaran N. The G protein family and their interaction with receptors. Endocrine Rev 1989; 10:317-331.
56. Birnbaumer L. Receptor-to-effector signaling through G proteins: roles for βγ dimers as well as α-subunits. Cell 1992; 71:1069-1072.
57. Simon MI, Strathman MP, Gautam N. Diversity of G proteins in signal transduction. Science 1991; 252:802-808.
58. Moxham CM, Hod Y, Malbon CC. Induction of Gαi2-specific antisense RNA in vivo inhibits neonatal growth. Science 1993; 260:991-965.
59. Wang H-Y, Watkins DC, Malbon CC. Antisense oligodeoxynucleotides to G_s protein α-subunit sequence accelerate differentiation of fibroblasts to adipocytes. Nature 1992; 358:334-337.
60. Lyons J, Landis CA, Harsh G et al. Two G protein oncogenes in human endocrine tumors. Science 1990; 249:655-659.
61. Van Noorden S, Polak JM, Pearse AGE. Single cellular origin of somatostatin and calcitonin in the rat thyroid gland. Histochemistry 1977; 53:243-247.
62. Deschenes RJ, Lorenz LJ, Haun RS et al. Cloning and sequence

analysis of a cDNA encoding rat preprocholecystokinin. Proc Natl Acad Sci USA 1984; 81:726-730.
63. Cremins JD, Michel J, Farah JM et al. Characterization of substance P-like immunoreactivity and tachykinin-encoding mRNAs in rat medullary thyroid carcinoma cell lines. J Neurochem 1992; 58:817-825.
64. Lindh B, Hokfelt T. Structural and functional aspects of acetylcholine peptide coexistence in the autonomic nervous system. Prog Brain Res 1990; 84:175-191.
65. Baetge G, Pintar JE, Gershon MD. Transiently catecholaminergic (TC) cells in the bowel of the fetal rat: precursors of noncatecholaminergic enteric neurons. Dev Biol 1990; 141:353-380.
66. deBustros A, Baylin SB, Levine MA et al. Cyclic AMP and phorbol esters separately induce growth inhibition, calcitonin secretion, and calcitonin gene transcription in cultured human medullary thyroid carcinoma. J Biol Chem 1986; 261:8036-8041.
67. Peleg S, Abruzzese RV, Cooper CW et al. Down regulation of calcitonin gene transcription by vitamin D requires two widely separated enhancer sequences. Mol Endocrinol 1993; 11:1750-1757.
68. Naveh-Many T, Silver J. Regulation of calcitonin gene transcription by vitamin D metabolites in vivo in the rat. J Clin Invest 1988; 81:1-14.
69. Lanigan TL, Tverberg LA, Russo AF. Retinoic acid repression of cell-specific helix-loop-helix-octamer activation of the calcitonin/calcitonin gene-related peptide enhancer. Molec Cell Biol 1993; 13:6079-6088.
70. Tverberg LA, Russo AF. Cell-specific glucocorticoid repression of calcitonin/calcitonin gene-related peptide transcription. J Biol Chem 1992; 267:17567-17573.
71. Tverberg LA, Russo AF. Regulation of the calcitonin/calcitonin gene-related peptide gene by cell-specific synergy between helix-loop-helix and octamer-binding transcription factors. J Biol Chem 1993; 268:15965-15973.
72. Ball DW, Compton D, Nelkin BD et al. Human calcitonin gene regulation by helix-loop-helix recognition sequences. Nucl Acids Res 1992; 20:117-123.
73. Peleg S, Abruzzese RV, Cote GJ et al. Transcription of the human calcitonin gene is mediated by a C cell-specific enhancer containing E-box-like elements. Mol Endocrinol 1990; 4:1750-1757.
74. DeBustros A, Lee RY, Compton D et al. Differential utilization of calcitonin gene regulatory DNA sequences in cultured lines of medullary thyroid carcinoma and small-cell lung carcinoma. Molec Cell Biol 1990; 10:1773-1778.
75. Ball DW, Azzoli CG, Baylin SB et al. Identification of a human achaete-scute homolog highly expressed in neuroendocrine tumors. Proc Natl Acad Sci USA 1993; 90:5648-5652.

76. Struhl K. Mechanisms for diversity in gene expression patterns. Neuron 1991; 7:177-181.
77. Beato M. Transcriptional control by nuclear receptors. FASEB J 1991; 5:2044-2051.
78. Kutoh E, Stromstedt PE, Poellinger L. Functional interference between the ubiquitous and constitutive octamer transcription factor 1 (OTF-1) and the glucocorticoid receptor by direct protein-protein interaction involving the homeo subdomain of OTF-1. Molec Cell Biol 1992; 12:4960-4969.
79. DeLuca LM. Retinoids and their receptors in differentiation, embryogenesis, and neoplasia. FASEB J 1991; 5:2924-2933.
80. Johnson JE, Birren SJ, Anderson DJ. Two rat homologues of Drosophila achaete-scute specifically expressed in neuronal precursors. Nature 1990; 346:858-861.
81. Lo L, Johnson JE, Wuenschell CW et al. Mammalian achaete-scute homolog 1 is transiently expressed by spatially restricted subsets of early neuroepithelial and neural crest cells. Genes and Devel 1991; 5:1524-1537.
82. Guillemot F, Lo L, Johnson J et al. Mammalian achaete-scute homologue-1 is required for the early development of olfactory and autonomic neurons. Cell 1993; 75:1-20.

CHAPTER 8

Progression of Medullary Thyroid Carcinoma

Barry D. Nelkin

INTRODUCTION

A universal aspect of cancer development is its multistage nature. Precancerous stages evolve to more advanced stages of malignancy and metastasis. This has been reviewed extensively by Foulds.[1,2] Based on observations of karyotypic progression in hematopoietic and other neoplasms, Nowell[3] proposed a clonal evolution model for tumor progression. In this model, variant cell clones which possess some growth advantage overgrow the other cells in the tumor.

In most cases, several events must accrue in a cell to produce a tumor, and several more may be required during tumor progression. These events effect changes in the phenotype of the cell due to changes in gene structure or expression; these changes may be genetic or epigenetic. While it is not certain how many of these changes are required for tumorigenesis, estimates have been attempted. Nordling[4] and Armitage and Doll[5] argued that the age-specific incidence rates of a cancer will be proportional to the (n-1) power of age, where n is the number of events required for carcinogenesis:

$$r = kt^{(n-1)}$$

Using published mortality data for several common tumor types, these authors estimated n as 6-7. This number is an estimate of the total number of events in tumor development and progression

Genetic Mechanisms in Multiple Endocrine Neoplasia Type 2A, edited by Barry D. Nelkin.

leading to the death of the patient. However, since it has been found that substantially fewer events may be required for development of a recognizable tumor (e.g., ref. 6), several subsequent steps must be required for tumor progression. Of cautionary importance, these are only very rough estimates of the number of events required for development of these stages of tumorigenesis. Among the factors contributing to this uncertainty are the inherent, probably untenable, assumptions that the probability of a given event does not change over time and that each event alone does not affect the growth of the cell population at risk. In addition, data on the age dependence of each stage of tumorigenesis are subject to artifacts such as ascertainment, detection and reporting biases. These difficulties have been discussed.[7,8]

The specific identity of most of these tumor progression events has not been well-characterized for most types of cancer. One notable exception is the extensive characterization of genetic and epigenetic changes in colon cancer development and progression, accomplished in large part by Vogelstein and colleagues.[9,10] In colon cancer, the characterized abnormalities include oncogene activation, tumor suppressor gene inactivation, DNA repair abnormalities and abnormalities in DNA methylation. Among these lesions, the abnormalities in DNA repair and methylation presumably are mechanisms leading to instability in structure and expression of other critical genes. Also, not every common abnormality in colon cancer is found in any given tumor; instead, it appears that the number of accrued abnormalities, rather than the specific abnormalities, results in progression of colon cancer. This somewhat surprising finding indicates that there are several pathways, for any tumor, that can lead to tumor progression. These studies have provided a model for the steps of tumor progression in other types of cancer, although the specific steps involved may differ among tumor types.

MTC affords an interesting opportunity to examine tumor progression, since several discrete accessible stages of the disease have been defined. In MEN 2 patients, the first recognizable abnormal stage is a widespread, polyclonal hyperplasia of the thyroid C-cells.[11] From this hyperplasia, one or more clonal microscopic carcinomas develop. These microscopic carcinomas eventually progress to palpably detectable carcinomas. In some cases, the relatively indolent carcinomas may become more aggressive, with poor prognosis for the patients. Application of the aforementioned equation for the number of events in multistage carcinogenesis[4,5] to age at onset data in MEN 2[12,13] suggests that several genetic or epigenetic events occur during MTC tumorigenesis (Fig. 8.1). In addition to the germline ret mutation, these data would suggest that 1-2 further events accrue during progression to the stage of C-cell hyperplasia detectable by the pentagastrin stimulation test (see chapter 1). Another 1-2 events may occur during progression from C-cell hyperplasia to clinical MTC. As discussed above, the exact number of events derived from this analysis may be inaccurate, but at very least the data are consistent with the multistage nature of MTC. In MEN 2, the identity of most of these changes is unknown. The only well-characterized specific genetic abnormality in MEN 2 is the mutation of the ret gene, as described in earlier chapters. Other changes in oncogene expression and possible tumor suppressor genes in MEN 2 will be discussed in chapter 9.

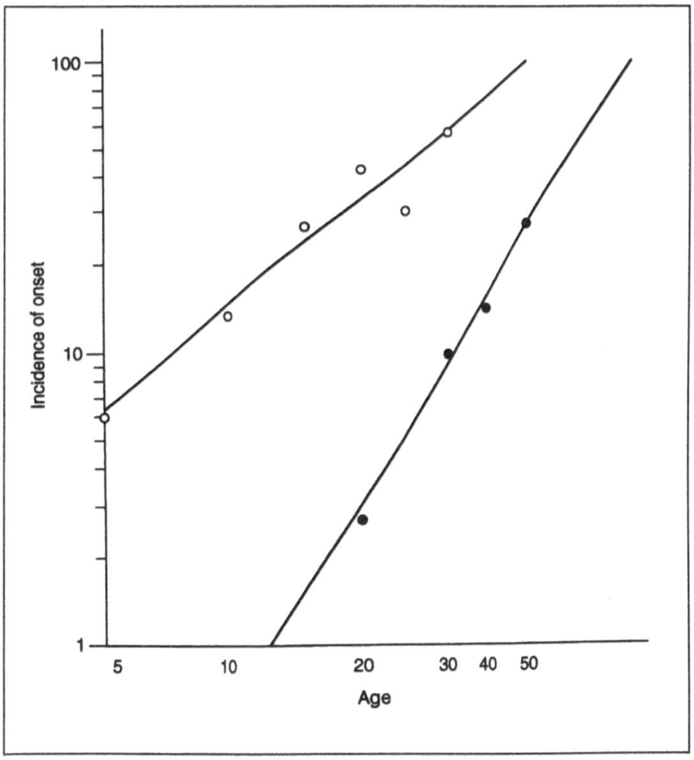

Fig. 8.1. Age-specific incidence (percent of at risk population) for onset of biochemically detectable C-cell hyperplasia (by positive pentagastrin stimulation test (°) or clinically evident MTC (•) in MEN 2A gene carriers. The steeper slope for onset of clinically evident MTC indicates that subsequent events are required for development of MTC from C-cell hyperplasia. Graphs were plotted from data in refs. 12 and 13.

There are some changes in gene expression which commonly take place during progression of MTC. These changes in gene expression parallel important changes in the phenotype of MTC. Some reported changes in gene expression in MTC are listed in Table 8.1. Even within this list, changes in some of the markers are controversial and have not been found in all studies. Moreover, the important unanswered question concerning all of these markers is which of these genes (if any) mediates tumor progression changes, and which of them only secondarily changes in response to the primary signals actually responsible for tumor progression. In the rest of this chapter, several of these changes and their possible relationship to MTC tumor progression will be discussed.

RET EXPRESSION AND THE DEVELOPMENT OF C-CELL HYPERPLASIA

Of the stages of tumor development in MEN 2, we apparently understand most about the transition from normal C-cells to C-cell

Table 8.1. Gene markers reported to change during MTC progression

Marker	Stage[a]	Direction	ref[b]
CT[c]	h,a	–	14-20
CGRP[c]	h	+	14,15
CEA	a	+	16
SRIF	s	–	21
N-myc[c]	s	+	22-24
CD15/leuMI	larger tumors, a	+	20,25,26
BRST-1	a	+	26
CGRP inducibility	a	+	27
PCNA	s	+	20,28
Polysialylated NCAM	h	+	29
Sialyl Le (a)	a	+	30
Secretogranin IV	c	–	31
Amyloid	s	–	32,33
TrkA, TrkC[c]	h,c,a	+	see text
TrkB[c]	h,c,a	–	see text
GRP	h	+	34
Bcl-2	s	–	35

[a] h, hyperplasia; c, carcinoma; a, aggressive tumor; s, survival
[b] includes both positive and negative reports
[c] discussed in text

hyperplasia. Since most of the thyroid C-cells in MEN 2 appear to undergo a polyclonal hyperplasia, it has been proposed that the germline activation of ret is sufficient to generate this initial mitogenic signal.[11,36-37] In turn, those markers which have been documented to change in hyperplastic C-cells (e.g., CGRP, trk family receptors, GRP, polysialylated NCAM; see Table 8.1) may do so as a response to the initial ret signal. As suggested above, there also may be subsequent changes necessary for progression of this initial C-cell hyperplasia to a biochemically detectable stage.

The ret gene is expressed in only a subset (10-25%) of normal thyroid C-cells.[38,39] If ret is not expressed in most normal C-cells, how can the ret mutation result in such widespread C-cell hyperplasia involving most of the C-cells? At least three models can be envisioned. In the first model, only those cells which express ret become hyperplastic; since these cells grow, while cells that do not express ret do not grow, it will appear that the majority of the population of C-cells is affected. In the second model, the C-cells that express ret induce mitogenesis in the C-cells that do not express ret, by a paracrine or endocrine mechanism. Ret expression may or may not be turned on in these cells. Since MTC, at the carcinoma stage, expresses ret, this model would suggest only those cells which actually express ret can progress from hyperplasia to carcinoma. In the third model, most or all of the thyroid C-cells in MEN 2 patients initially express ret unlike the heterogeneous expression seen in normal thyroids. In this model, one may envision that normal C-cell precursors express ret, and terminal differentiation of C-cells involves the extinction of ret gene expression. The constitutive activation of ret in MEN 2 may block this terminal differentiation, thus allowing both continued ret expression and cell proliferation. Resolution of these models awaits examination of ret expression in situ in C-cell hyperplasia.

CHANGES IN CALCITONIN GENE EXPRESSION DURING MTC PROGRESSION

Calcitonin (CT) is the most extensively examined phenotypic marker in medullary thyroid carcinoma. The CT gene expression pattern undergoes changes throughout MTC tumor progression. This section presents a review of the control of CT gene expression and examines the abnormalities in CT gene expression that occur in MTC. While two calcitonin genes have been identified in humans and other species,[40,41] this discussion will focus on CALC-I (hereafter called the CT gene), the gene that encodes the CT and calcitonin gene-related peptide (CGRP) species that have been most extensively characterized.

Changes in Calcitonin Gene Alternative RNA Processing

CT gene expression is highly tissue specific; expression is primarily limited to the thyroid cells, central and peripheral nervous systems, anterior pituitary, lung endocrine cells and pancreatic islet cells. The CT gene is composed of six exons (Fig. 8.2). From a single species of initial CT gene RNA transcript, two major mature mRNAs are generated by alternative RNA processing.[40-42] One mRNA, which contains sequences from exons 1, 2, 3 and 4, encodes CT. The other mRNA, which contains sequences from exons 1, 2, 3, 5 and 6, encodes CGRP. The alternative RNA processing choice is tightly controlled in normal cells expressing the CT gene. In normal thyroid C-cells, about 99% of the CT gene transcripts are processed to CT mRNA; in other tissues, virtually 100% of the CT gene transcripts are processed to CGRP mRNA.[44] Thus, these cells must regulate splicing of the common exon 3 splice donor to either the CT-specific exon 4 or the CGRP-specific exon 5 splice acceptor.

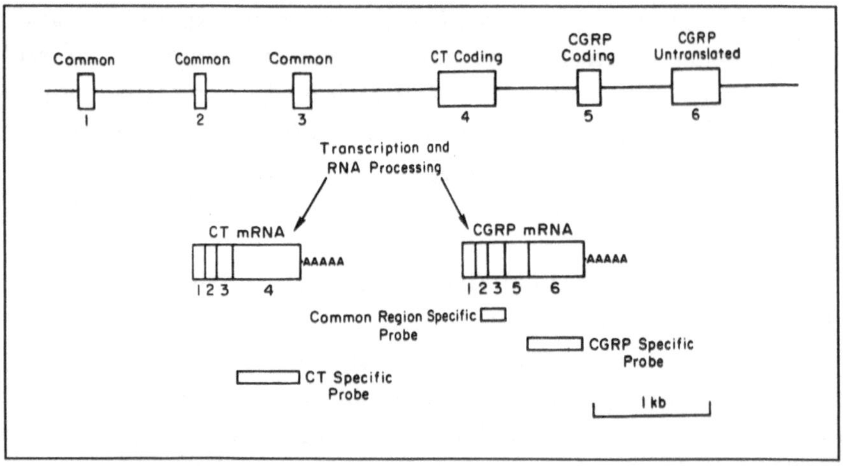

Fig. 8.2. Structure of the human CALC-I gene, and CT and CGRP mRNAs. The mechanism for RNA processing in this gene is discussed in the text. From ref. 77, with permission of the American Association for Cancer Research.

In MTC development, as early as the C-cell hyperplasia stage, the tight control of CT gene RNA processing is relaxed, and both CT and CGRP are produced.[14,15,43,44] Since both CT and CGRP mRNAs are produced from a single initial RNA precursor, the control of this mRNA processing choice must be exerted at the level of trans-acting RNA processing machinery. This regulatory machinery is likely to be involved in alternative RNA processing choices in many genes. Therefore, the changes seen in CT/CGRP expression in MTC probably reflect a more global change in alternative RNA processing choices. Such changes could have pleiotropic effects on the cell phenotype. Therefore, it is important to consider the mechanism of control of alternative RNA processing of the CT gene, since this mechanism may control aspects of the changing phenotype of MTC.

What specifies the highly regulated control of this alternative splice choice in normal cells? This question was initially explored in a transgenic mouse model system.[45] These transgenic animals, in which the calcitonin gene was expressed under the control of the ubiquitously active mouse metallothionein promoter, processed the initial RNA transcript predominantly to CT mRNA in most tissues, except in heart and brain, in which both CT and CGRP mRNAs were found. Further examination of the brains of these animals, using in situ hybridization and immunohistochemistry, demonstrated that the CT and CGRP mRNA splice patterns were cell-specific; CGRP was predominantly found in neurons and CT was found predominantly in non-neuronal cells. These experiments showed that most cells have the machinery to determine an unambiguous processing choice for CT gene transcripts. In addition, these experiments suggested that the "default" processing choice, employed by most cells, was to CT mRNA, and only neurons had additional factors required for processing to CGRP mRNA; however, alternative models, that only neurons lacked the factors required for processing to CT mRNA, or that specific factors were needed for each processing pathway, were not excluded by these experiments. Subsequent studies have attempted to define the cis- and trans-factors which mediate the CT gene splice choice, using transfection in tissue culture and in vitro splicing assays. Important for the development of these model systems was the finding that tissue culture cells, most of which do not express the endogenous CT gene, can also tightly regulate processing of a transfected

CT gene; for example, HeLa and 293 cells were found to process the initial CT gene transcript to CT mRNA, and F9 cells were found to process the transcript to CGRP mRNA.[46-49]

Two general models for control of CT gene RNA processing were originally envisioned,[50] based on the possible mechanisms of control of eukaryotic RNA processing (for a review of the biochemistry of the splicing reaction, see ref. 51). Minimally, splicing of eukaryotic RNA transcripts requires interaction of the RNA processing machinery (spliceosome) with the 3' end of the upstream exon (splice donor), the 5' end of the downstream exon (splice acceptor) and the splice branchpoint. Other regions of the gene, including exon sequences and the polyA addition site, may also influence splicing. In the CT gene, the splice choice could be dictated either by use of the alternative polyA addition sites at the ends of exon 4 and exon 6 (which would then specify the splicing choice by default), or the choice of splice acceptor site might be the primary mechanism. These alternative possible mechanisms have been extensively explored by mutagenesis of the CT gene. If the splice choice were dictated by polyA addition site usage, then one would predict that the splice choice could be modified by deleting or altering the exon 4 polyA addition site.[47] This proved not to be the case. Neither HeLa (CT specific) nor F9 (CGRP specific) cells exhibited altered splicing when transfected with constructs lacking a functional exon 4 polyA addition site.[46,47] Thus, it appears that the processing choice is made primarily by choice of the splice acceptor site.

The control of splice acceptor choice in the CT gene has been studied extensively in vitro, using in vitro splicing or transient transfection assays. Mixing of HeLa and F9 extracts in an in vitro splicing reaction resulted in production of CGRP mRNA, demonstrating that splicing to CGRP is dominant over splicing to CT.[52] Deletion of the exon 4 splice acceptor allowed HeLa cells, which normally spliced to CT, to splice to CGRP. However, deletion of the exon 5 splice acceptor, or even all of exons 5 and 6, did not result in production of CT mRNA in cells which normally splice to CGRP; instead, the precursor RNA remained unspliced.[47,53] Together with the results of the transgenic mouse experiments discussed above, these experiments suggested that CGRP-producing cells express a negative regulatory factor which interacts with the exon 4 splice acceptor, allowing the skipping of exon 4. Using

RNA from the exon 4 splice acceptor site as a probe, Cote et al[53] have detected a 66 kDa protein-RNA complex formed with HeLa cell (CT-specific splicing) nuclear extracts, but not with F9 cell (CGRP specific splicing) nuclear extracts. Mutations in the RNA, which resulted in loss of splice acceptor activity in HeLa cells, also resulted in loss of protein binding.[54] A mixture of HeLa and F9 nuclear extracts did not form this protein-RNA complex. This suggests that the 66 kDa protein may be a positive regulatory factor which recognizes the exon 4 splice acceptor, and that a negative regulatory factor in F9 cells blocks this recognition. Fractionation of nuclear extracts from rat brain has yielded a 43 kDa protein which can inhibit utilization of the exon 4 splice acceptor in an in vitro splicing reaction.[55] This 43 kDa protein, which has not yet been identified, is a fascinating candidate to mediate CT gene alternative splicing.

However, the control of CT gene alternative splicing may be even more complicated. Further mutational analyses have identified at least three domains of exon 4 which augment CT-specific splicing in HeLa or 293 cells.[53,56] Extensive mutation of another exon 4 domain, from base 40 to base 56 of exon 4, allowed CT-specific splicing in F9 cells.[56] In addition, sequences within intron 3 can affect splicing specificity.[47-49,58-59] The most dramatic alterations appear to result from mutagenesis of sequences around the putative lariat branchpoint. In most higher eukaryotic genes, lariat branching occurs at a highly conserved adenosine within the degenerate consensus sequence YNYURAY. However, in intron 3 of the CT gene, this consensus is violated, so that the putative branchpoint is C (rat) or U (human) instead of A. Conversion of this base to A, to match the metazoan branchpoint consensus, allowed splicing specifically to CT mRNA in F9 cells.[48,59] Conversely, in Hela cells, deletion of this putative branchpoint and surrounding region, from base -58 to -17 upstream from exon 4,[47] or mutagenesis of all the adenosines in this region,[57] allowed splicing to CGRP mRNA. These experiments also suggested that HeLa cells may be able to employ alternative branchpoints in intron 3. Taken together, these data indicate that several domains together are important for the decision to splice to CT or CGRP mRNA. These data remain consistent with a model in which a trans-acting negative regulatory factor in CGRP-producing cells interacts with the initial CT transcript, allowing skipping of exon 4. Such a factor would probably undergo a complex interaction with sequences near the lariat branchpoint, the splice acceptor and within exon 4. Such an interaction could depend upon RNA secondary structure of binding of a protein or ribonucleoprotein complex. An alternative model suggests that the alternative splicing choice in the CT gene may depend upon recognition, by cell specific trans-acting factors, of a combination of suboptimal splicing sequences in the CT gene, and that the final splicing choice depends upon competition among factors which favor processing to CT or CGRP mRNA.[57]

Decreased CT During Tumor Progression

As indicated in Table 8.1, decreased CT is an indicator of poor prognosis of human MTC. Trump et al[60] initially observed a patient with widely metastatic MTC, who exhibited remarkably low plasma calcitonin levels; these levels did not correlate with estimated tumor size. Immunohistochemistry of the primary tumor showed cellular heterogeneity for CT; some cells stained intensely for CT, while other cells did not stain. The metastases did not stain for CT. This pattern was common in the small subset of patients who died of aggressive MTC, and loss of CT immunostaining has been shown in several studies to be a marker of poor prognosis in MTC.[16-19]

What is the mechanism of loss of CT in these aggressive tumors? One can envision several possible control levels for CT expression. These include transcription, mRNA splicing, mRNA stability, translation, peptide stability and export. In human MTC, it has been found that CT and CGRP levels are correlated.[19] This suggests that changes in alternative splicing do not account for the loss of immunohistochemically detected CT in aggressive MTC. Several studies have indicated that even in MTC tumors which have little CT by immunohistochemistry, CT mRNA, as detected by in situ hybridization, is undiminished.[14,61] This would suggest that the control of CT peptide expression is at a step subsequent to RNA metabolism, i.e., in translation or a subsequent step. Two groups have speculated that the loss of intracellular CT may be due to a defect in intracellular peptide storage.[14,60] This would appear to be unlikely, however, since it would predict that plasma levels of CT would remain high or even rise in aggressive cases of MTC; instead, several reports have shown that plasma levels of CT are diminished in parallel with tumor CT levels.[18,26,60,62] Taken together, these data suggest that the loss of CT immunostaining in aggressive MTC may be due to a defect in translation or protein stability.

EXPRESSION OF N-MYC AND TRK RECEPTOR FAMILY MEMBERS IN MTC–A PARALLEL WITH NEUROBLASTOMA?

In this section, I discuss the speculative possibility that N-myc and trk family neurotrophin receptors are involved in MTC tumor progression. This is based on consideration of expression patterns in MTC, and comparison with another neural crest derived tumor, neuroblastoma, in which N-myc and trk family gene expression are important prognostic indicators.

N-MYC AND TRK FAMILY GENE EXPRESSION IN NEUROBLASTOMA

N-myc is an important prognostic indicator in neuroblastoma. Overexpression of N-myc, which usually, but not always, results from gene amplification, correlates well with poor prognosis of neuroblastoma.[63]

As mentioned earlier in this volume (chapter 4), three members of the trk family of neurotrophin receptors, trkA, trkB and trkC, are now known. The ligands for these receptors are the neurotrophins NGF, BDNF (as well as NT-4/5) and NT-3, respectively. The downstream signaling pathways activated by the trk receptors were discussed in Chapter 4. In neuroblastomas, trkA expression is found preferentially in those tumors with good prognosis.[64] In contrast, trkB is also often expressed in neuroblastomas which exhibit poor prognosis; since most neuroblastomas also express BDNF, the ligand for trkB, this suggests the possibility of autocrine stimulation.[65]

The biological effects of expression of these genes in neuroblastoma have been explored extensively in cell culture. Overexpression of N-myc in neuroblastoma cell lines resulted in more aggressive behavior (Peng, personal communication). Antisense-mediated abrogation of N-myc expression resulted in slowed growth and reduced cloning efficiency.[66,67] Together, these results indicate that N-myc expression is important for growth control in neuroblastoma cells.

Introduction of a trkA gene into the HTLA 230 neuroblastoma cell line led to NGF-inducible cell differentiation and growth arrest.[68] In many neuroblastoma cell lines derived from aggressive N-myc amplified tumors, the trkA signaling pathway is impaired.[69] Activation of trkB by BDNF addition in neuroblastoma cell lines also resulted in cell differentiation; however, this was not accompanied by slower growth, and in one cell line, growth was actually stimulated by BDNF.[64,70] These results suggest that trkA and trkB may have opposing effects on growth in neuroblastoma; trkA may act to inhibit growth and trkB may stimulate growth. Preliminary evidence suggests that trkA and trkB may act, in part, by modulation of N-myc expression. TrkA induced cell differentiation and growth arrest was accompanied by down regulation of N-myc.[71] In addition, in the SMS-KCN neuroblastoma cell line, which expresses both trkA and trkB, NGF treatment decreased N-myc expression, while BDNF treatment increased N-myc expression.[70]

Expression of N-myc and trk Family Genes in MTC

N-myc expression has been observed by RNase protection in situ hybridization and immunohistochemistry in many cases of MTC.[22-24] Expression was not detected in normal C-cells. The tumors which were found to express N-myc were larger than those which did not express N-myc, and N-myc expression was positively correlated with shorter survival time. N-myc expression in MTC was heterogeneous within tumors. In some tumors, the highest expression appeared to be in areas of poor differentiation, suggesting a possible inverse correlation between N-myc expression and MTC cell differentiation. However, N-myc expression in MTC was not associated with gene amplification, as has been found commonly in neuroblastoma.

All three members of the trk receptor family are expressed in MTC, and expression of these genes is progressively altered during MTC progression (McGregor et al, unpublished). In normal adult thyroid C-cells, only trkB could be detected by immunohistochemistry. The trkB expression was heterogeneous; 11% to 28% of calcitonin-positive C-cells were trkB positive. In C-cell hyperplasia, most of the C-cells were positive for trkB. At this stage, trkA and trkC were also detectable by immunohistochemistry in some cases. Further changes in trk receptor family expression were evident in the carcinomas. The strong trkB expression, which were common in earlier stages, was rarely seen, while strong trkC expression were usually seen. Moderate or strong trkA expression was also seen in about 60% of MTC tumors.

In light of the data on neuroblastoma, one may speculate that in MTC different trk family members may also serve different functions. In normal thyroid, trkB may mark those C-cells which retain proliferative capacity; this would be consistent with the finding that hyperplastic C-cells express high levels of trkB. The expression of trkA or trkC in later stages of MTC suggests that these receptors may influence tumor progression.

MTC DIFFERENTIATION IN CULTURE—REVERSAL OF TUMOR PROGRESSION?

This chapter has discussed several candidate genes and processes which may be involved in tumor progression. It would be useful to have an MTC model for tumor progression or differentiation, in which to examine and modulate the expression of these genes. MTC cells have been shown to undergo partial differentiation in response to a number of agents, including protein kinase A and C agonists, glucocorticoids and bFGF (reviewed in Chapter 7; ref. 37). We have employed a human MTC cell line, TT, to study MTC differentiation. The TT cell line was derived from an apparently sporadic case of MTC.[72] It expresses one wild-type and one mutant (cys634arg) ret allele.[73] TT cells produce roughly equal amounts of CT and CGRP mRNAs.

We have observed an extensive differentiation program in TT cells in response to activation of a ras signal transduction pathway. When TT cells were infected by HaMSV, they exhibited increases in both transcription of the CT gene, and in their CT/CGRP mRNA ratio.[74] The cells also altered their morphology (cell rounding), produced more neurosecretory granules and substantially slowed their growth. Some of these changes, especially the switch toward restoration of control of CT-specific splicing and the slowed cell growth, suggest the possibility that v-ras^{H} induced a reversal of tumor progression in TT cells in vitro. Since abnormalities in CT mRNA processing occur early in MTC development in C-cell hyperplasia, this suggests that, at least in some respects, ras activation may allow TT cells to revert to resemble normal C-cells.

As described in chapter 4, ras signal transduction in many cell types has been shown to involve activation of the c-raf protein kinase. Many of the effects of ras require c-raf activation as an obligatory step; however, c-raf-independent ras signal transduction pathways have been reported. In order to examine whether the differentiation effects we observed in ras-infected TT cells can be recapitulated by raf, we have used a construct[75] in which a transforming portion of the human c-raf-1 gene was fused to the hormone binding domain of the estrogen receptor gene. In this construct, activation of the c-raf-1 gene is estrogen-dependent. Upon addition of estradiol, these TT: Raf-1ER cells exhibited the same changes we had previously seen upon ras infection.[76] In these cells, we also observed complete loss of expression of ret concomitant with raf-induced differentiation. Since ret is the initial stimulus for C-cell hyperplasia in MEN 2, one may speculate that ret may be the main target of the ras-raf signal transduction pathway in TT cells, and that loss of the ret mitogenic signal may be responsible for the differentiation program we have observed.

The ras/raf signal transduction pathway may be involved in normal C-cell differentiation; it has been shown previously that normal C-cells express high levels of ras protein.[78,79] In this model, the ras/raf signal transduction pathway may be responsible for down regulation of ret during terminal differentiation of normal C-cells during development. C-cells which express mutant ret alleles, as in MEN 2, may block these ras/raf derived terminal differentiation signals leading to the widespread C-cell hyperplasia seen in MEN 2. Our data on TT cells may suggest that constitutively activated ras or raf may provide a sufficiently strong signal to compete with ret, allowing normal differentiation and cessation of proliferation. The MTC cell culture model should allow an assessment of the effects of other genes, such as N-myc and trk receptor family members, on MTC progression and differentiation.

References

1. Foulds L. The experimental study of tumor progression: a review. Cancer Res 1954; 14:317-39.
2. Foulds L. Neoplastic Development. New York: Academic Press 1969, 1975.
3. Nowell PC. The clonal evolution of tumor cell populations. Science (Washington DC) 1976; 194:23-8.
4. Nordling CO. A new theory on the cancer-inducing mechanism.

Br J Cancer 1953; 7:68- 72.
5. Armitage P, Doll R. The age distribution of cancer and a multistage theory of carcinogenesis. Br J Cancer 1954; 8:1-12.
6. Ashley DJ. On the incidence of carcinoma of the prostate. J Pathol Bacteriol 1965; 90:217-24.
7. Whittemore AS. Quantitative theories of oncogenesis. Adv Canc Res 1978; 27:55-88.
8. Muir CS, Fraumeni JF Jr, Doll R. The interpretation of time trends. Cancer Surveys 1994; 19/20:5-21.
9. Fearon ER, Vogelstein B. A genetic model for colorectal tumorigenesis. Cell 1990; 61:759-67.
10. Vogelstein B, Kinzler KW. The multistep nature of cancer. Trends in Genetics 1993; 9:138-41.
11. Wolfe HJ, Melvin KEW, Cervi-Skinner SJ et al. C-cell hyperplasia preceding medullary thyroid carcinoma. N Engl J Med 1973; 289:437-41.
12. Gagel RF, Jackson CE, Block MA et al. Age-related probability of development of hereditary medullary thyroid carcinoma. J Pediatr 1982; 101:941-6.
13. Ponder, BAJ, Coffey R, Gagel RF et al. Risk estimation and screening in families of patients with medullary thyroid carcinoma. Lancet 1988; 1:397-400.
14. Boultwood J, Wynford-Thomas D, Richards GP et al. In-situ analysis of calcitonin and CGRP expression in medullary thyroid carcinoma. Clin Endocrinol 1990; 33:381-90.
15. Williams ED, Ponder BAJ, Craig RK. Immunohistochemical study of calcitonin gene- related peptide in human medullary carcinoma and C cell hyperplasia. Clin Endocrinol 1987; 27:107-14.
16. Lippman SM, Mendelsohn G, Trump DL et al. The prognostic and biologic significance of cellular heterogeneity in medullary thyroid carcinoma: A study of calcitonin, L-dopa decarboxylase, and histaminase. J Clin Endocrinol Metab 1982; 54:233-40.
17. Saad MF, Ordonez NG, Guido JJ et al. The prognostic value of immunostaining in medullary carcinoma of the thyroid. J Clin Endocrinol Metab 1984; 59:850-6.
18. Bergholm U, Adami H-O, Auer G et al. Histopathologic characteristics and nuclear DNA content as prognostic factors in medullary thyroid carcinoma. Cancer 1989; 64:135-42.
19. Takami H, Bessho T, Kameya T et al. Immunohistochemical study of medullary thyroid carcinoma: Relationship of clinical features to prognostic factors in 36 patients. World J Surg 1988; 12:572-9.
20. Riddell DA, Lampe HB, Cramer H et al. Medullary thyroid carcinoma: prognostic factors. J Otolaryngol 1993; 22:180-3.
21. Pacini F, Basolo F, Ekisei R et al. Medullary thyroid cancer. An immunohistochemical and humoral study using six separate antigens. Am J Clin Pathol 1991; 95:300-8.

22. Roncalli M, Viale G, Grimelius L et al. Prognostic value of N-myc immunoreactivity in medullary thyroid carcinoma. Cancer 1994; 74:134-41.
23. Boultwood J, Wyllie FS, Williams ED et al. N-myc expression in neoplasia of human thyroid C-cells. Cancer Res 1988; 48:4073-7.
24. Klimpfinger M, Ruhri C, Putz B et al. Oncogene expression in a medullary thyroid carcinoma. Virchows Arch B Cell Pathol Incl Mol Pathol 1988; 54:256-9.
25. Neuhold N, Langle F, Gnant M et al. Relationship of CD15 immunoreactivity and prognosis in sporadic medullary thyroid carcinoma. J Cancer Res Clin Oncol 1992; 118:629-34.
26. Langle F, Soliman T, Neuhold N et al. CD15 (LeuM1) immunoreactivity: Prognostic factor for sporadic and hereditary medullary thyroid cancer? Study Group on Multiple Endocrine Neoplasia of Austria. World J Surg 1994; 18:583-7.
27. Takami H, Ito K. Calcitonin gene-related peptide as a tumor marker for medullary thyroid carcinoma. Int Surg 1992; 77:181-5.
28. Skopelitou A, Korkolopoulou P, Papanikolaou A et al. Proliferating cell nuclear antigen (PCNA) in medullary thyroid carcinoma. J Cancer Res Clin Oncol 1993; 119:379-81.
29. Komminoth P, Roth J, Saremaslani P et al. Polysialic acid of the neural cell adhesion molecule in the human thyroid: A marker for medullary thyroid carcinoma and primary C-cell hyperplasia. An immunohistochemical study on 79 thyroid lesions. Am J Surg Pathol 1994; 18:399-411.
30. Vierbuchen M, Schroder S, Larene A et al. Native and sialic acid masked (a) antigen reactivity in medullary thyroid carcinoma. Distinct tumor-associated and prognostic relevant antigens. Virchows Archiv A Pathol Anat Histopathol 1994; 424:205-11.
31. Neuhold N, Ullrich R. Secretogranin IV immunoreactivity in medullary thyroid carcinoma: An immunohistochemical study of 62 cases. Virchows Arch A Pathol Anat Histopathol 1993; 423:85-9.
32. Harach HR, Wilander E, Grimelius L et al. Chromogranin A immunoreactivity compared with argyrophilia, calcitonin immunoreactivity, and amyloid as tumour markers in the histopathological diagnosis of medullary (C-cell) thyroid carcinoma. Pathol Res Pract 1992; 188:123-30.
33. Pyke CM, Hay ID, Goellner JR et al. Prognostic significance of calcitonin immunoreactivity, amyloid staining, and flow cytometric DNA measurements in medullary thyroid carcinoma. Surgery 1991; 110:964-70.
34. Sunday ME, Wolfe HJ, Roos BA et al. Gastrin-releasing peptide gene expression in developing, hyperplastic and neoplastic thyroid C-cells. Endocrinol 1988; 122:1551-8.
35. Viale G, Roncalli M, Grimelius L et al. Prognostic value of BCL-2 immunoreactivity in medullary thyroid carcinoma. Human Path

1995; 26:945-50.
36. Nelkin BD, Nakamura Y, White RW et al. Low incidence of loss of chromosome 10 in sporadic and hereditary human medullary thyroid carcinoma. Cancer Res 1989; 49:4114-9.
37. Nelkin BD, Ball DW, Baylin SB. Molecular abnormalities in tumors associated with multiple endocrine neoplasia, type 2. Endocrinol Metab Clin North Amer 1994; 23:187-213.
38. Fabien N, Paulin C, Santoro M et al. Expression of the RET protooncogene in normal human C-cells and adrenal medulla. Int J Onc 1994; 4:623-6.
39. Tsuzuki T, Takahashi M, Asai N et al. Spatial and temporal expression of the ret proto-oncogene product in embryonic, infant and adult rat tissues. Oncogene 1995; 1005-17.
40. Höppener JWM. The human calcitonin genes. Ph.D. thesis. University of Utrecht. Utrecht, Netherlands 1988;37-75.
41. Rosenfeld MG, Emeson RB, Yeakley JM et al. Calcitonin generelated peptide: A neuropeptide generated as a consequence of tissue-specific, developmentally regulated alternative RNA processing events. Ann NY Acad Sci 1992; 657:1-17.
42. Amara SG, Jonas V, Rosenfeld MG et al. Alternative RNA processing generates mRNAs encoding diffeent polypeptide products. Nature 1982; 298:240-4.
43. Rosenfeld MG, Mermod J-J, Amara SG et al. Production of a novel neuropeptide encoded by the calcitonin gene via tissue-specific RNA processing. Nature 1983; 304:129-35.
44. Sabate, MI, Stolarsky, LS, Polak JM et al. Regulation of neuroendocrine gene expression by alternative RNA processing. J Biol Chem 1985; 260:2589-92.
45. Crenshaw EB III, Russo AF, Swanson LW et al. Neuron-specific alternative RNA processing in transgenic mice expressing a metallothionein-calcitonin fusion gene. Cell 1987; 49:389-98.
46. Leff SE, Evans RM, Rosenfeld MG. Splice commitment dictates neuron-specific alternative RNA processing in calcitonin/CGRP gene expression. Cell 1987; 48:517-24.
47. Emeson RB, Hedjran R, Yeakley JM et al. Alternative production of calcitonin and CGRP mRNA is regulated at the calcitonin-specific splice acceptor. Nature (London) 1989; 341:76-80.
48. Bovenberg RAL, Adema GJ, Baas PD. Model for tissue specific calcitonin/CGRP-I RNA processing from in vitro experiments. Nucleic Acids Res 1988; 16:7867-83.
49. Adema GJ, Bovenberg RAL, Baas PD. Unusual branch point selection involved in splicing of the alternatively processed calcitonin/CGRP-I pre-mRNA. Nucleic Acids Res 1988; 16:9513-26.
50. Amara SG, Evans RM, Rosenfeld MG. Calcitonin/calcitonin generelated peptide transcription unit: tissue-specific expression involves selective use of alternative polyadenylation sites. Mol Cell Biol 1984;

4:2151-60.
51. Moore MJ, Query CC, Sharp PA. Splicing of precursors to mRNAs by the spliceosome. In: Gesteland RF, Atkins JF, eds. The RNA World. Cold Spring Harbor: Cold Spring Harbor Laboratory Press, 1993:303-357.
52. Cote GJ, Nguyen IN, Lips CJM et al. Validation of an in vitro RNA processing system for CT/CGRP precursor mRNA. Nucleic Acids Res 1991; 19:3601-6.
53. Cote GJ, Stolow, DT, Peleg S et al. Identification of exon sequences and an exon binding protein involved in alternative RNA splicing of calcitonin/CGRP. Nucleic Acids Res 1992; 20:2361-6.
54. Cote GJ. Alternative RNA splicing of calcitonin/calcitonin gene-related peptide minigene transcripts in a thyroid C-cell line. Biochem Biophys Res Commun 1994; 200:993-8.
55. Roesser JR, Litschwager K, Leff SE. Regulation of tissue-specific splicing of the calcitonin/CGRP gene by RNA binding proteins. J. Biol. Chem. 1993; 268:8366-75.
56. van Oers CCM, Adema GJ, Zandberg H et al. Two different sequence elements within exon 4 are necessary for calcitonin-specific splicing of the human calcitonin/calcitonin gene-related peptide I pre-mRNA. Mol Cell Biol 1994; 14:951-960.
57. Yeakley JM, Hedjran F, Morfin J-P et al. Control of calcitonin/ calcitonin gene-related paptide pre-mRNA processing by constitutive intron and exon elements. Mol Cell Biol 1993; 13:5999-6011.
58. Adema GJ, van Hulst KL, Baas PD. Uridine branch acceptor is a cis-acting element involved in regulation of the alternative processing of calcitonin/CGRP-I pre-mRNA. Nucleic Acids Res 1990; 18:5365-73
59. Adema GJ, Baas PD. Deregulation of alternative processing of calcitonin/CGRP-I pre-mRNA by a single point mutation. Biochem Biophys Res Commun 1991; 178:985-92.
60. Trump DL, Mendelsohn G, Baylin SB. Discordance between plasma calcitonin and tumor-cell mass in medullary thyroid carcinoma. New Engl J Med 1979; 301:253-5.
61. Le Guellec P, Dumas S, Volle GE et al. An efficient method to detect calcitonin mRNA in normal and neoplastic rat C-cells (medullary thyroid carcinoma) by in situ hybridization using a digoxigenin-labeled synthetic oligodeoxyribonucleotide probe. J Histochem Cytochem 1993; 41:389-95.
62. Saad MF, Ordonez NG, Rashid RK et al. Medullary carcinoma of the thyroid. A study of the clinical features and prognostic factors in 161 patients. Medicine 1984; 63:319-42.
63. Brodeur GM, Seeger RC, Schwab M et al. Amplification of N-myc in untreated human neuroblastomas correlates with advanced disease stage. Science 1984; 224:1121-4.
64. Nakagawara A, Arima-Nakagawara M, Scavarda NJ et al. Associa-

tion between high levels of expression of the TRK gene and favorable outcome in human neuroblastoma. N Engl J Med 1993; 328:847-54.
65. Nakagawara A, Azar CG, Scavarda NJ et al. Expression and function of TRK-B and BDNF in human neuroblastomas. Mol Cell Biol 1994; 14:759-67.
66. Larcher JC, Basseville M, Vayssiere JL et al. Growth inhibition of N1E-115 mouse neuroblastoma cells by c-myc or N-myc antisense oligodeoxynucleotides causes limited differentiation but is not coupled to neurite formation. Biochem Biophys Res Commun 1992; 185:915-24.
67. Schmidt ML, Salwen HR, Manohar CF et al. The biological effects of antisense N-myc expression in human neuroblastoma. 1994; 5: 171-8.
68. Matsushima H, Bogenmann E. Expression of trkA cDNA in neuroblastomas mediates differentiation in vitro and in vivo. Mol Cell Biol 1993; 13:7447-56.
69. Azar CG, Scavarda NJ, Reynolds CP et al. Multiple defects of the nerve growth factor receptor in human neuroblastomas. Cell Growth Differ 1990; 1:421-8.
70. Matsumoto K, Wada RK, Yamashiro JM et al. Constitutive N-myc gene expression inhibits trkA mediated neuronal differentiation. Oncogene 1995; 10:1915-25.
71. Thiele C, Matsumoto K, Lucarelli E et al. Brain derived neurotrophic factor (BDNF) stimulates NMYC transcription in human neuroblastoma cells. Proc Amer Assoc for Cancer Res 1995; 36:559.
72. Leong SS, Horoszewicz JS, Shimaoka K, et al. A new cell line for study of human medullary thyroid carcinoma. In: Andreoli M, Monaco F, Robbins J, eds. Advances in Thyroid Neoplasia. Rome, Italy: Field Eductional Italia 1981; 95-108.
73. Carlomagno F, Salvatore D, Santoro M. Expression of brain-derived neurotrophic factor and p145TrkB affects survival, differentiation, and invasiveness of human neuroblastoma cells. Cancer Res 1995; 55:1798-806.
74. Nakagawa T, Mabry M, de Bustros A et al. Introduction of Harvey v-ras oncogene induces differentiation of cultured human medullary thyroid carcinoma cells. Proc Natl Acad Sci USA 1987; 84:5923-7.
75. Samuels ML, Weber MJ, Bishop JM et al. Conditional transformation of cells and rapid activation of the mitogen-activated protein kinase cascade by an estradiol-dependent human raf-1 protein kinase. Mol Cell Biol 1993; 13:6241-52.
76. Carson EB, McMahon M, Baylin SB et al. Ret gene silencing is associated with raf-1-induced medullary thyroid carcinoma cell differentiation. Cancer Res 1995; 55:2048-52.

77. Nelkin BD, Chen KY, de Bustros A et al. Changes in calcitonin gene RNA processing during growth of a human medullary thyroid carcinoma cell line. Cancer Res 1989; 49:6949-52.
78. Chesa PG, Rettig WJ, Melamed MR et al. Expression of p21ras in normal and malignant human tissues: lack of association with proliferation and malignancy. Proc Natl Acad Sci 1987; 84:3234-8.
79. Furth ME, Aldrich TH, Cordon-Cardo C. Expression of ras protooncogene proteins in normal human tissues. Oncogene 1987; 1:47-58.

CHAPTER 9

Molecular Events in the Development and Progression of Medullary Thyroid Cancer and Pheochromocytoma

Jeffrey F. Moley

INTRODUCTION

Alterations in several classes of genes can contribute to the development and progression of tumors. These abnormalities can be classified into three general groups: dominant oncogenes, tumor suppressor genes, and genetic instability genes. This chapter reviews what is known about abnormalities in these classes of genes in MTC and pheochromocytoma.

DOMINANT ONCOGENES

Transforming genes whose protein products have a positive or dominant effect on cell growth and proliferative capacity are referred to as dominant oncogenes. The dominant acting oncogenes can be classified into three categories based on their function. In the first category are the molecules which act as growth factors or growth factor receptors. Oncogenes which act by this pathway may

Genetic Mechanisms in Multiple Endocrine Neoplasia Type 2, edited by Barry D. Nelkin.

stimulate cell proliferation and transformation by causing stimulation or disruption of signal transduction pathways. The second category includes the second messengers and signal transduction molecules themselves, which act in the cytoplasm and the inner surface of the cell membrane. This group of molecules includes the ras gene products. The third category of dominant oncogenes includes genes which encode nuclear proteins and transcription factors which are involved in controlling gene expression. The dominant oncogene role of ret in the development of pheos and MTCs is beginning to be understood and was discussed in earlier chapters. The subsequent events which result in tumor progression are only beginning to be investigated. When considering cancers in general, cases such as hereditary retinoblastoma, where abnormalities of a single gene result in the development of a tumor, are probably very unusual. A germ-line ret gene mutation alone does not appear to be sufficient for development of pheochromocytomas, as they occur in only 40% of patients with MEN 2A and 2B. When they do arise, examination of the remainder of the gland reveals areas of normal appearing adrenal medulla, regions of medullary hyperplasia and occasionally, multifocal pheochromocytomas. These observations suggest that the constitutional ret gene mutation, which is present in all cells of the body, predisposes adrenal medullary cells to molecular events which result in the progression to hyperplasia and neoplasia. In other types of cancer, including lung, colon and breast cancer, multiple genetic abnormalities are found in tumor cells, and it is likely that these tumors arise after multiple defects in different genes have occurred. The contribution of multiple genetic events to the development of neoplasia has been most comprehensively described in colorectal neoplasia by Vogelstein and collaborators.[1] The genetic analysis of MTCs and pheos has not been exhaustive, and particularly in the case of MTCs, very few abnormalities aside from RET mutations have been reported.

Absence of amplification of N-myc, c-myc and ErbB2 has been reported in MTCs and pheos.[2,3] Roncalli et al reported that N-myc expression in greater than 10% of MTC tumor cells, as detected by immunohistochemistry, was associated with poorer survival, sporadic disease and male sex.[3] They found no evidence of gene amplification and did not determine the basis for the overexpression. The possible role of N-myc expression in MTC was discussed in the previous chapter. Our group reported absence of mutation of the H-ras, N-ras and K-ras genes in a series of pheos and MTCs analyzed by direct sequencing.[4] Likewise, examination of nerve growth factor and nerve growth factor receptor (p75) showed no abnormality at the DNA or RNA levels.[2]

TUMOR SUPPRESSOR GENES

In normal cells, tumor suppressor genes participate in the control of cell growth.[1] Loss of these genes, by deletion, mutation or loss of function, can remove normal cellular growth control. Cavenee et al showed that loss of function of a tumor suppressor gene is often associated with physical loss, by recombination, nondisjunction or deletion, of a normal allele of the gene.[5] This loss can be detected by loss of heterozygous markers in the region of the tumor suppressor gene; loss of heterozygosity (LOH) in a specific genomic region at high frequency in a tumor-type can be considered evidence for the presence of a tumor suppressor gene involved in development of that tumor. Depending upon the mechanism of loss of chromosomal material, LOH can affect the entire chromosome, one arm of the chromosome, or a smaller portion of DNA. The shortest region of overlap (SRO) of LOH, among several tumors of a given type, can serve to delineate the location of the tumor suppressor gene, facilitating its subsequent identification through positional cloning.

Since only the tumor DNA will exhibit LOH, it is important to minimize the contribution by normal cells to the DNA sample, or the LOH can be masked. Moreover, since LOH can be easier to demonstrate with some probes than with others, normal cell contamination can lead to erroneous conclusions regarding the SRO; this, in turn, can lead to errors in mapping the putative tumor suppressor genes. Tumor cells are often admixed with normal cells and stroma; in some instances, the non-tumor cells will constitute the majority of the sample. Microdissection of the tumor prior to DNA extraction can reduce the proportion of normal cell DNA in tumor DNA samples, allowing reliable conclusions to be drawn.

There have been several studies evaluating loss of heterozygosity at tumor suppressor loci in pheos and MTCs, which are summarized in Table 9.1. The cumulative data indicate a higher than background incidence of LOH in pheos on chromosome arms 1p,

3p, 17p and 22q. In MTCs, at least two studies suggest a significant incidence of 1p LOH;[6,7] evaluation of other chromosomal arms yielded no consistent findings of LOH. Lack of significant LOH on 10q at the ret locus supports the hypothesis that the ret proto-oncogene acts as a dominant oncogene as opposed to a tumor suppressor gene.[8-10]

For the most part, these LOH studies have not yet identified candidate tumor suppressor genes in MEN 2. LOH analysis on 1p in pheochromocytomas suggests a very large region of deletion.[11] Shin et al[12] have suggested that the SRO for LOH on chromosome 1p is between D1S63 (1pter-1p36.1) and D1S73 (1p22-p13), a region encompassing most of 1p. This LOH was found both in sporadic pheos and in those from patients with MEN 2. Our recent studies,[13] and those of Mulligan et al,[7] have indicated that the entire short arm of chromosome 1 is lost in pheos from patients with MEN 2A and 2B. Fine mapping of the region of deletion suggests a possible common breakpoint in the centromeric region defined by the markers D1S514 and D1S442.[13] We have also found that allelic loss on 1p is not influenced by the parent of origin, and that 1p LOH can be demonstrated in regions of adrenal medullary hyperplasia from patients with MEN 2A and 2B, suggesting that 1p LOH is an early event in tumorigenesis. The high rate of LOH on 3p in pheochromocytomas suggests an as yet undefined tumor suppressor locus.[15] LOH on 17p suggests the possible involvement of the p53 gene.[16] Existing reports on p53 mutations in pheochromocytomas are conflicting. While two Japanese groups have reported no evidence of p53 mutations in pheos,[17,18] a Chinese group reported p53 mutations in 5/6 tumors tested.[19] Four of these mutations were in exon 4. Our group has found no evidence of p53 mutations in exons 4-10 by SSCP analysis performed on 20 MTCs and 20 pheos. While more studies will be necessary to resolve this conflict, the data may suggest the possibility of geographical differences in the involvement of p53 in MTC and pheochromocytoma. There are other cancer types which consistently exhibit LOH on chromosomes 1p, 3p and 17p (not including p53). It remains unknown whether the LOH seen in pheos targets the same tumor suppressor genes affected in these other cancers.

PHEOCHROMOCYTOMAS IN OTHER TUMOR PREDISPOSING SYNDROMES

Pheochromocytomas occur in von Hippel-Lindau (VHL) disease and neurofibromatosis type 1 (NF1), both of which are caused by mutations in tumor suppressor genes.

VHL

The VHL gene is located on chromosome 3p.[20] A clue to the possible mechanism of disease development in VHL has been reported recently.[21-24] The VHL protein binds the B and C subunits of elongin, a protein which promotes transcriptional elongation. Binding of elongin BC by normal VHL interferes with elongin function, allowing a block in transcriptional elongation. Several genes associated with cell proliferations are controlled at the level of RNA elongation, including genes of the myc family and c-fos. Loss of VHL function may lead to overexpression of these genes, resulting in increased cell proliferation.

Interestingly, pheochromocytoma only occurs in 20% of VHL families.[25] In these cases, the VHL protein usually contains a missense mutation (often at codon 238); in those VHL families that do not have pheochromocytomas, the mutation is more often a nonsense codon, a frameshift or a deletion.[25] This suggests that the development of pheochromocytoma in VHL disease is augmented by full length VHL protein; however, the mechanism by which VHL protein acts in pheochromocytoma development is unknown.

Mutational analysis of the VHL gene in non-VHL pheochromocytoma has not been completed. SSCP analysis of a small number of sporadic pheos did not identify any mutations in the VHL gene.[26] However, this analysis did not examine the complete VHL gene, and the VHL gene remains a candidate for a tumor suppressor gene in non-VHL pheochromocytoma.

NF1

The neurofibromin tumor suppressor gene, which is mutated in the NF1 syndrome, is located on chromosome 17q. Tumors, including pheochromocytomas, from NF1 patients exhibit loss of neurofibromin protein.

We reported decreased or absent expression of the neurofibromin gene in 7/20 pheochromocytomas from patients with MEN 2 and sporadic disease.[27] Because neurofibromin is ubiquitously expressed, its lack of expression indicates that neurofibromin may play a role in the development or progression of pheochromocytomas from patients who do not have NF1. Because of the extremely large size of the neurofibromin gene, mutational analysis has not yet been feasible. The amino acid sequence of neurofibromin is related to that of p120GAP, a protein which stimulates the GTPase activity of ras.[28] p120GAP activity results in ras activity being down-regulated, by driving ras into the inactive ras-GDP form. Neurofibromin has been shown biochemically to have GAP activity.[29] Thus, one mechanism by which neurofibromin may work, as a tumor suppressor, is by controlling the function of ras. Loss of neurofibromin would be equivalent to activation of ras. However, this attractive model for neurofibromin function has two serious problems. First, not all tumors which have lost neurofibromin have high levels of ras-GTP. Second, in several tumor types, including pheochromocytoma, ras activation results in differentiation and growth arrest.

To resolve these apparent difficulties in neurofibromin action, several interesting models have been proposed. These models provide novel approaches to possible mechanisms of tumor suppression by neurofibromin, and the models can be tested experimentally. Li et al[30] noted that p120GAP not only has GTPase stimulating activity, but also has an effector function downstream from ras; neurofibromin may also have an effector function. In cell types in which ras is growth promoting, loss of neurofibromin may augment growth by increasing ras-GTP; in cells, such as adrenal chromaffin cells, in which ras is growth inhibiting, loss of neurofibromin may interfere with ras activity by removing an effector function.

Viskochil et al[31] have proposed a model in which ras delivers a neutral activating signal to the effectors neurofibromin and p120GAP. In this model, neurofibromin signals the cells to differentiate, while p120GAP signals the cells to grow. The final effect of ras on any cell type would depend upon competition between neurofibromin and p120GAP. In cells with high concentrations of neurofibromin, ras activation would result in differentiation, while cells with high concentrations of p120GAP will grow and may become transformed. Loss of neurofibromin would result in cell transformation, since ras signaling would proceed through the p120GAP effector.

McCormick[32] has suggested that the result of a ras signal may depend upon dosage. In PC12 cells, transient ras activation appears to result in cell growth, while sustained ras activation appears to result in cell differentiation.[33] Similarly, loss of neurofibromin may provide modest activation of ras function, resulting in cell growth even in those cells in which strong ras activation would result in differentiation.

DNA REPAIR GENES

Deficiencies in excision repair or mismatch repair have been reported in several cancer predisposition syndromes.[34,35] Such deficiencies result in an increased mutation frequency, presumably leading to mutation in as yet unspecified oncogenes or tumor suppressor genes. Mismatch repair deficiencies, most commonly associated with hereditary nonpolyposis colon cancer, result from mutation in one of at least four genes (MLH1, MSH2, PMS1, PMS2).[35,36] Inactivation of the mismatch repair genes result in tumors with genomic instability, evidenced by alterations in microsatellite sequences.

There have been no reports of the involvement of DNA repair genes (MLH1, MSH2) in development or progression of pheos or MTCs, and replication error of repeats have not been a consistent finding in these tumors (unpublished data).

SUMMARY AND CONCLUSIONS

While the molecular analysis of pheos and MTCs has not been exhaustive, several conclusions can be made. First of all, pheos and MTCs appear to arise by different mechanisms and demonstrate very different biological behavior. Few abnormalities have been reported in MTCs, and because of the complete penetrance of this trait, it is likely that few molecular events in addition to the constitutional ret mutation are needed for tumorigenesis in the hereditary form of this disease. In sporadic MTC, it is likely that somatic ret mutations play an important role in tumor development. In the future, there will probably be more reports of ret mutations in these tumors, particularly in areas of the gene not affected by the MEN 2A and 2B mutations.

Pheochromocytomas arise in only 40% of patients with MEN 2A and 2B, and they are also a component of other hereditary cancer syndromes, VHL and NF1. It is likely, then, that these tumors may arise by several different mechanisms. We found that LOH of chromosome 1p is always present in pheos from patients with MEN2A and 2B. This finding appears to be present in very early stages of the tumor as well, and it is our contention that 1p LOH is likely a necessary event for the formation of pheos in patients with MEN 2A and MEN 2B. 1p LOH is not found with the same consistency in sporadic pheos, though it is found with a high enough frequency to lead to the conclusion that 1p LOH plays a role in the development of some sporadic pheos. Ret mutations have been reported to occur in sporadic pheos, and a more detailed evaluation of the entire gene may demonstrate that there is a higher incidence of ret mutations in sporadic pheos than we now know. LOH analysis has demonstrated several regions of the genome where tumor suppressor genes may be inactivated in these tumors, which is in contrast to the lack of such findings in MTC, and this also lends support to the conclusion that development of pheos in both hereditary and sporadic settings may occur through a variety of pathways.

References

1. Fearon ER, Vogelstein B. A genetic model for colorectal tumorigenesis. Cell 1990; 61:759-67.
2. Moley J, Wallin G, Brother M et al. Oncogene and growth factor expression in MEN related tumors. Henry Ford Hospital Medical Journal 1992; 40:284-8.
3. Roncalli M, Viale G, Grimelius L et al. Prognostic value of N-myc immunoreactivity in medullary thyroid carcinoma. Cancer 1994; 74:134-41.
4. Moley JF, Brother MB, Wells SA Jr et al. Low frequency of ras gene mutations in neuroblastomas, pheochromocytomas, and medullary thyroid cancers. Cancer Research 1991; 51:1596-9.
5. Cavenee WK, Dryja TP, Phillips RA et al. Expression of recessive alleles by chromosomal mechanisms in retinoblastoma. Nature 1983; 305:779-84.
6. Mathew CGP, Smith BA, Thorpe K et al. Deletion of genes on chromosome 1 in endocrine neoplasia. Nature 1987; 328: 524-6.
7. Mulligan LM, Gardner E, Smith BA et al. Genetic events in tumour initiation and progression in multiple endocrine neoplasia type 2. Genes, Chromosomes and Cancer 1993; 6:166-7.

8. Landsvater RM, Mathew CG, Smith BA et al. Development of multiple endocrine neoplasia type 2A does not involve substantial deletions of chromosome 10. Genomics 1989; 4:246-50.
9. Nelkin BD, Nakamura Y, White RW et al. Low incidence of loss of chromosome 10 in sporadic and hereditary human medullary thyroid carcinoma. Cancer Research 1989; 49:4114-9.
10. Khosla S, Patel VM, Hay ID et al. Loss of heterozygosity suggests multiple genetic alterations in pheochromocytomas and medullary thyroid carcinomas. Journal of Clinical Investigation 1991; 87:1691-9.
11. Moley J, Brother M, Fong C et al. Consistent association of 1p loss of heterozygosity with pheochromocytomas from patients with multiple endocrine neoplasia type 2 syndromes. Cancer Research 1992; 52:770-4.
12. Shin E, Fujita S, Takami K et al. Deletion mapping of chromosome 1p and 22q in pheochromocytoma. Japanese Journal of Cancer Research 1993; 84:402-8.
13. Moley JF, Marshall HN. 1p Deletions in human pheochromocytomas share a common pericentromeric breakpoint and do not involve imprinting. American Journal of Human Genetics 1994; 55:A347.
14. Moley JF, Marshall MN, Gagliardi G et al. Loss of heterozygosity of 1p is an early event in the development of pheochromocytomas from patients with multiple endocrine neoplasia types 2A and 2B. Proceedings of the American Association for Cancer Research 1995; 36:A3265.
15. Dou S, Toshima K, Liu L et al. Identification of chromosomal loci for tumor suppressor loci implicated in progression of pheochromocytoma and medullary thyroid carcinoma. American Journal of Human Genetics 1994; 55:A20.
16. Lum S, Brodeur G, Wells S et al. Loss of heterozygosity at the p53 (17p13) locus in human pheochromocytomas. Proceedings of the American Society of Human Genetics 1992: A244.
17. Yoshimoto K, Iwahana H, Fukuda A et al. Role of p53 mutations in endocrine tumorigenesis: mutation detection by polymerase chain reaction-single strand conformation polymorphism. Cancer Research 1992; 52:5061-4.
18. Yana I, Nakamura T, Shin E et al. Inactivation of the p53 gene is not required for tumorigenesis of medullary thyroid carcinoma or pheochromocytoma. Japanese Journal of Cancer Research 1992; 83:1113-6.
19. Lin SR, Lee YJ, Tsai JH. Mutations of the p53 gene in human functional adrenal neoplasms. Journal of Clinical Endocrinolology and Metabolism 1994; 78:483-91.
20. Latif F, Tory K, Gnarra J et al. Identification of the von Hippel-Lindau disease tumor suppressor gene. Science 1993; 260:1317-20.

21. Duan DR, Pause A, Burgess WH et al. Inhibition of transcription elongation by the VHL tumor suppressor protein. Science 1995; 269:1402-6.
22. Aso T, Lane WS, Conaway JW, Conaway RC. Elongin (SIII): A multisubunit regulator of elongation by RNA polymerase II. Science 1995; 269:1439-43.
23. Kibel A, Iliopoulos O, DeCaprio JA et al. Binding of the von Hippel-Lindau tumor suppressor protein to elongin B and C. Science 1995; 269:1444-6.
24. Krumm A, Groudine M. Tumor suppression and transcription elongation: the dire consequences of changing partners. Science 1995; 269:1400-1.
25. Crossey PA, Richards FM, Foster K et al. Identification of intragenic mutations in the von Hippel-Lindau disease tumor suppressor gene and correlation with disease phenotype. Human Mol Genetics 1994; 3:1303-8.
26. Gnarra JR, Tory K, Weng Y et al. Mutations of the VHL tumor suppressor gene in renal carcinoma. Nature Genetics 1994; 7:85-90.
27. Gutmann DH, Geist RT, Rose K et al. Loss of neurofibromatosis type I (NF1) gene expression in pheochromocytomas from patients without NF1. Genes, Chromosomes and Cancer 1995; 13:104-9.
28. Xu G, O'Connell P, Viskochil D et al. The neurofibromatosis type 1 gene encodes a protein related to GAP. Cell 1990; 62:599-608.
29. Xu G, Lin B, Tanaka K et al. The catalytic domain of the neurofibromatosis type 1 gene product stimulates ras GTPase and complements ira mutants of S. cerevisiae. Cell 1990; 63:835-41.
30. Li Y, Bollag G, Clark R et al. Somatic mutations in the neurofibromatosis 1 gene in human tumors. Cell 1992; 69:275-81.
31. Viskochil D, White R, Cawthon R. The neurofibromatosis type 1 gene. Annu Rev Neurosci 1993; 16:183-205.
32. McCormick F. Ras signaling and NF1. Current Opinion in Genetics and Development 1995; 5:51-5.
33. Traverse S, Gomez N, Paterson H et al. Sustained activation of the mitogen-activated protein (MAP) kinase cascade may be required for differentiation of PC12 cells. Comparison of the effects of nerve growth factor and epidermal growth factor. Biochem J 1992; 288:351-5.
34. Bootsma D, Weeda G, Vermeulen W et al. Nucleotide excision repair syndromes: molecular basis and clinical symptoms. Philos Trans R Soc Lond B Biol Sci 1995; 347:75-81.
35. Kolodner RD, Hall NR, Lipford J et al. Human mismatch repair genes and their association with hereditary non-polyposis colon cancer. Cold Spring Harb Symp Quant Biol 1994; 59:331-8.
36. Liu B, Parsons R, Papadopoulos N et al. Analysis of mismatch repair genes in hereditary non-polyposis colorectal cancer patients. Nature Medicine 1996; 2:169-74.

37. Yang KP, Nguyen CV, Castillo SG et al. Deletion mapping on the distal third region of chromosome 1p in multiple endocrine neoplasia type IIA. Anticancer Research 1990; 10:527-33.

CHAPTER 10

Conclusion

The preceding chapters establish a framework to answer basic questions about the biology of MEN 2; the resulting knowledge will undoubtedly provide new therapeutic targets for MTC, pheochromocytoma and biologically related cancers. Here, I would like to point out some of the questions which have been raised in the text, and which seem to me to be of special interest for future study. Of course, these choices reflect my bias; and others might emphasize alternate lines of investigation.

1. In MTC and pheochromocytoma, it seems clear that an unusual cell response to common signal transduction pathways occurs (chapters 3, 4 and 7-9). In both cell types, activation of the ras/raf/MAPK pathway results in differentiation; in many other cell types, activation of this pathway is associated with a mitogenic response. Other neural or neuroendocrine tumors also differentiate when this pathway is activated.[1-3] What in these cells specifies the differentiation response? An answer to this question could provide new targets for the common human cancers in which ras activation frequently occurs. A similar question regards cell cycle control. In MTC and pheochromocytoma, inactivation of the p53 tumor suppressor gene, one of the most common abnormalities in human cancer, is rarely seen. This suggests that p53 inactivation may not provide a growth advantage in MTC or pheochromocytoma, and raises the question of why this is so.

2. What are the downstream effectors of ret? Some of the possible signal transduction pathways have been discussed in chapter 3, but the characterization of these pathways is just beginning. It appears likely that the signal transduction pathways will differ among cells, and between MEN 2A and MEN 2B.

---

3. How can the same mutations in ret result in MEN 2A, including pheochromocytomas, in some families, and FMTC, lacking pheochromocytomas, in others (chapters 1 and 2)? It is possible that some of the FMTC families will eventually be shown to be MEN 2A families, but in a number of cases, the FMTC families are large enough to assure a true FMTC phenotype. This suggests the possibility that a modifier locus, determining disease phenotype, may segregate in these families. A conceptually similar locus has been identified for murine colon cancer.[4] Possibly, an animal model for such a modifier locus in MEN 2 already exists. Lines of transgenic mice expressing c-mos develop MTC, pheochromocytoma, or both tumors; the tumor developed is dependent both upon the transgenic strain and the genetic background.[5]

4. What is the function of ret in neural crest development? Many neural crest cells express ret. However, the only documented abnormalities in ret knockout mice are in the kidney and the enteric nervous system.[6] In MEN 2, which exhibits constitutive activation of ret, abnormalities are also limited to a subset of neural crest derived cells. Is ret expression redundant in some neural crest cells? Is it expressed in an inactive form, either due to alternative splicing[7] or failure to interact with signal transduction pathways? An obvious corollary question is what is the ligand for ret and where is it expressed; in addition to its implications for developmental biology, this may be important for possible autocrine or paracrine stimulation in ret-expressing tumors.

5. What are the biological bases and implications of the phenotypic heterogeneity observed in both normal adrenal chromaffin cells and normal C-cells (chapters 5 and 8). Normal adrenal chromaffin cells are heterogeneous in their expression of several neuronal or chromaffin markers; similarly, C-cells are heterogeneous in their expression of ret and trkB. Does this heterogeneity indicate a phenotypic difference, for example, in their commitment to terminal differentiation or in their proliferation capability? This may be especially important in determining the mechanism by which ret can induce hyperplasia in these cells, and in interpreting the species differences in the incidence of adrenal hyperplasia and pheochromocytoma (chapter 6).

6. What is the developmental origin of the thyroid C-cell? These cells appear to originate in the neural crest and migrate to the thyroid via the lowest pharyngeal pouch.[8] However, interme-

diate steps in C-cell differentiation are not described in comparable detail to the sympathoadrenal lineage of adrenal medullary cells (chapter 5). Knowledge of these steps might allow intervention in MTC by restoration of normal differentiation pathways. A related question concerns the origin of the rare mixed medullary-follicular and medullary-papillary cancers. These cancers, which exhibit markers of both lineages in a single cell population, have been proposed to derive from either a bipotential stem cell, a collision of two independent tumors, or ectopic expression of tumor markers.[9] To interpret these cases, it is necessary to know the extent of overlap of differential potential of the stem cells which gives rise to the normal thyroid cell lineages.

7. What are the steps in MTC and pheochromocytoma progression? The discussion of these events in chapters 8 and 9 identified several prime suspects, but convicted no culprits. It will be important to examine the effect of altering these genes in tissue culture. It will also be important to develop transgenic models for MEN 2, using activated ret genes; this work is already in progress in several laboratories. Then, the effects of altering expression of genes such as N-myc, trk family members, NF1 and VHL can be observed in vivo, by crossing transgenic strains.

It is clear that the research which has advanced our knowledge of the biology of MEN 2, and which is already having dramatic impact on the clinical management of the disease, has its roots in basic biological research unrelated to MEN 2. While this statement is obvious to the readers of this book, it must be emphasized to those who support us that basic biology underlies medical progress. The remarkable advances in such research have been threatened recently by funding crises. We have a duty to thank those scientist-administrators and public policymakers who have, through brilliance and great effort, augmented the biomedical research effort in the United States. Among these, I point out Dr. Harold Varmus, Director, National Institutes of Health, and Dr. Richard Klausner, Director, National Cancer Institute, who have reworked the NIH grant system to elicit further efficiency from existing funds. I also point out Sen. Mark Hatfield (R-Ore.) and Rep. John Porter (R-Ill.), who in Congress have championed the U.S. government's support of biomedical research. The insight of these and others like them will lead to improved healthcare well into the next century.

References

1. Mabry M, Nakagawa T, Nelkin BD et al. Viral Harvey ras oncogene insertion: a model for tumor progression of human small cell lung cancer. Proc Natl Acad Sci USA 1988; 85:6523-7.
2. Conrad KE, Gutierrez-Hartmann A. The *ras* and protein kinase A pathways are mutually antagonistic in regulating rat prolactin promoter activity. Oncogene 1992; 7:1279-86.
3. Celano P, Berchtold CM, Mabry M et al. Induction of markers of normal differentiation in human colon carcinoma cells by the v-ras^H oncogene. Cell Growth Diff 1993; 4:341- 7.
4. Dietrich WF, Lander ES, Smith JS et al. Genetic identification of Mom-1, a major modifier locus affecting *Min*-induced intestinal neoplasia in the mouse. Cell 1993; 75: 631-9.
5. Schulz N, Propst F, Rosenberg MP et al. Pheochromocytomas and C-cell thyroid neoplasms in transgenic c-mos mice: A model for the human multiple endocrine neoplasia type 2 syndrome. Cancer Res 1992; 52:450-5.
6. Schuchardt A, D'Agati V, Larsson-Blomberg L et al. Defects in the kidney and enteric nervous system of mice lacking the tyrosine kinase receptor Ret. Nature 1994; 367:380-3.
7. Lorenzo MJ, Eng C, Mulligan LM et al. Multiple mRNA isoforms of the human RET proto-oncogene generated by alternate splicing. Oncogene 1995; 10:1377-83.
8. Fontaine J. Multistep migration of calcitonin cell precursors during ontogeny of the mouse pharynx. Gen Comp Endocrin 1979; 37:81-92.
9. DeLellis RA. The pathology of medullary thyroid carcinoma and its precursors. In: LiVolsi VA, DeLellis RA, eds. Pathology of the Parathyroid and Thyroid Glands. Baltimore: Williams and Wilkins, 1993:72-102.

Note added in proof: The ligand for ret has been identified as GDNF (Trupp et al, Nature 1996; 381:785-9. Durbec et al, Nature 1996; 381:789-93). The complex of ret and GDNF includes a third component, GDNFR-α (Jing et al, Cell 1996; 85:1113-24. Treanor et al, Nature 1996; 382:80-3). The importance of GDNF and GDNFR-α in development of normal thyroid C-cells, adrenal medulla, MTC and pheochromocytoma has not yet been elucidated.

INDEX

T

V